Reservoir engineering: guidelines for practice

Edward M Gosschalk

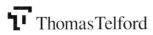
ThomasTelford

Published by Thomas Telford Publishing, Thomas Telford Ltd, 1 Heron Quay, London E14 4JD.
URL: http://www.thomastelford.com

Distributors for Thomas Telford books are
USA: ASCE Press, 1801 Alexander Bell Drive, Reston, VA 20191-4400, USA
Japan: Maruzen Co. Ltd, Book Department, 3–10 Nihonbashi 2-chome, Chuo-ku, Tokyo 103
Australia: DA Books and Journals, 648 Whitehorse Road, Mitcham 3132, Victoria

First published 2002

Also available from Thomas Telford Books

Hydraulic gates and valves in free surface flow and submerged outlets, 2nd edition J. Lewin.
ISBN 07277 2990X

Silt and siltation problems and solutions, edited by I. Jefferson, M. Rosenbaum and I. Smalley.
ISBN 07277 28865

River diversions, edited by D. Ramsbottom. ISBN 07277 29594

Dynamics of estuarine muds, R.J.S. Whitehouse, R.L. Soulsby, W. Roberts and H.J. Mitchner.
ISBN 07277 28644

A catalogue record for this book is available from the British Library

ISBN: 0 7277 3099 1

© E.M. Gosschalk and Thomas Telford Limited 2001

All rights, including translation, reserved. Except as permitted by the Copyright, Designs and Patents Act 1988, no part of this publication may be reproduced, stored in a retrieval system or transmitted in any form or by any means, electronic, mechanical, photocopying or otherwise, without the prior written permission of the Publishing Director, Thomas Telford Publishing, Thomas Telford Ltd, 1 Heron Quay, London E14 4JD.

This book is published on the understanding that the author is solely responsible for the statements made and opinions expressed in it and that its publication does not necessarily imply that such statements and/or opinions are or reflect the views or opinions of the publishers. While every effort has been made to ensure that the statements made and the opinions expressed in this publication provide a safe and accurate guide, no liability or responsibility can be accepted in this respect by the author or publishers.

Typeset by Academic & Technical, Bristol
Printed and bound in Great Britain by MPG Books Ltd, Bodmin, Cornwall

Contents

Preface		ix
Acknowledgements		xi

Part I. Conceptual and planning practice for reservoirs — 1

Chapter 1. Introduction and philosophy of approach — 3
 Introduction — 3
 Background — 3
 Disciplines needed — 4
 Constraints on training and experience — 5
 The Engineering thread — 6
 Mode of work — 8
 The need for mutual understanding — 9
 The roles of empirical and theoretical methods — 10
 References — 13

Chapter 2. Objectives — 15
 The purposes of reservoirs — 15
 The engineering planning of reservoirs — 17
 Plan of work — 19
 Reference — 21

Chapter 3. Selection of potential dam sites and conceptual schemes — 23
 Appraisal of data and formulation of possible projects — 23
 Prefeasibility studies — 29
 Choice of dam type — 35
 Choice of size of dam — 42
 References — 44

Chapter 4.	**Investigation of selected sites and geological studies**	**45**
	Site investigations	45
	Methods	47
	Conclusion	61
	References	61
Chapter 5.	**Hydraulic studies**	**63**
	Introduction	63
	Past experience	64
	Analytical methods	66
	Computational models	67
	Physical models	71
	Bibliography	81
Chapter 6.	**Hydrological studies**	**83**
	Background	83
	Deterministic methods	84
	Flood studies	86
	Droughts	97
	References	99
	Addendum to Chapter 6: Guidelines from the United States of America	100
Chapter 7.	**Spillways**	**101**
	Intensity of flood flow to be provided for	101
	The effect of type of dam on type of spillway needed	105
	Energy dispersal	111
	References	115
	Further reading	116
Chapter 8.	**River diversions during construction**	**117**
	Intensity of flood discharge to be provided for	117
	Methods of river diversion	118
	Closure of river diversions on completion	120
Chapter 9.	**Seismic loading**	**123**
	The relevance of seismic risk to dams in the UK	123
	The Engineering Guide to Seismic Risk to Dams in the United Kingdom	128
	Peak ground accelerations for design in relation to factors of safety and accelerations liable to be suffered	131

Case histories of seismic incidents	136
Conclusions from case histories on improving the earthquake resistance of dams	159
Conclusion	161
References	162
Further reading	163

Part II. Development practice for reservoirs — 165

Chapter 10. Water conduits for reservoirs — 167
Purposes	167
Types	167
Routing	174
Permissible velocities of flow	178
Tunnel lining	179
Steel lining in tunnels	181
Grouting in tunnels	182
Final tunnel cross section	182
References	183

Chapter 11. Tunnelling problems and excavation of shafts — 185
Low stresses in rock surround	185
Two more case histories	191
Methods of excavating shafts	195
Concluding remarks on tunnelling problems	197
Reference	198

Chapter 12. Electro-mechanical equipment and controls — 199
Introductory remarks	199
Turbines	200
Generators	207
Transformers	207
Main travelling cranes	208
Transmission systems	210
Hydraulic valves and gates	211
References	218

Chapter 13. Environmental considerations — 219
Introduction	219
Overview of the arguments for more large reservoir projects	223
Dams and development — the World Commission on Dams	226

	Environmental impact assessments	229
	References	238

Chapter 14. Costs and benefits — **241**
    General — 241
    Cost estimates — 242
    Principles of project appraisal — 247
    References — 254

Chapter 15. Efficient management for irrigation — **255**
    Introduction — 255
    The meaning of efficiency — 256
    Planning for efficiency and control — 258
    Basic criteria — 261
    Water requirements — 261
    Water distribution systems — 263
    Balancing storages — 264
    Sources of water supply — 265
    Analysis — 266
    Control — 266
    Forecasting — 267
    Planned water shortages — 268
    Final design and economic analysis — 269
    References — 269

Chapter 16. Small hydropower — **271**
    Introduction — 271
    Criteria for successful development — 272
    Sound planning and design — 273
    Multiple development — 274
    Operation at optimum load factor — 275
    Conjunctive operation — 276
    The needs for small hydropower — 276
    Different users — 279
    Classifying and selecting small hydro schemes — 283
    Conclusions — 287
    Further information — 288
    References — 288

Chapter 17. Safety and inspection of reservoirs — **289**
    Introduction — 289
    The Reservoirs Act 1975 — 291
    Proposed changes — 298

	Reviews by panels of experts	299
	Remedial works	300
	References	303

Chapter 18.	**Operation and maintenance, monitoring and inspection**	**305**
	General	305
	Final construction report	305
	Operating and maintenance instructions	306
	Monitoring and inspection	308
	Training and staffing	310
	Potential unforeseen human, mechanical and electronic problems	311
	Investigation and solution of leakage problems observed during inspections	313
	Other problems	315
	References	316

Index 327

Preface

Edward Gosschalk is a Partner of The Gosschalk Partnership. He was previously head of the Dams and Hydropower Division of Sir William Halcrow & Partners, Consulting Civil Engineers. Before that he was head of Halcrow's Technical Services Division, responsible for geotechnical and hydrological services, etc. He is a member of the All Reservoirs Panel under the 1975 Reservoirs Safety Act. He was a member of the Working Party appointed to review the Institution of Civil Engineers' *Engineering Guide to Floods and Reservoir Safety* and took part in the preparation of the revised edition, published in 1996. He is one of the principal authors of *An Engineering Guide to Seismic Risk to Dams in the United Kingdom*, sponsored by the Department of the Environment.

The Author has specialized in dams and hydropower and utilization of water resources for most of his working life. He has visited numerous overseas countries and worked for long periods in Ghana, Malaysia (on reservoirs of the Muda Irrigation Project) and Sri Lanka (where he was Resident Director on the Kotmale Hydropower Project on the Mahaweli Ganga). He was Chief Dams/Hydroelectric Power Engineer for the preparation of master plans in Ethiopia and China. He was joint Project Director for a hydrological survey for Halcrow of the Irrawaddy River Delta in Burma (now Myanmar). From 1980 until 1983 he was an external consultant and supervisor for the Irrigation MSc course at Southampton University. More recently he has been a specialist adviser on hydropower projects, including mini-hydro, in China, Indonesia and the Philippines. He is the author of more than 30 papers and publications.

The Author intends that this book should convey his key thoughts and knowledge of reservoir engineering practice, which could take an engineer years of involvement in reservoir engineering to acquire. The emphasis is on important aspects that are not readily available from text books, published work or lectures. Thus the book is not a comprehensive manual but it provides guidelines that are supplementary to a conventional higher education in engineering and are especially aimed at providing essential understanding to those aspiring to hold or actually holding responsibilities at senior level in the wide field of reservoir engineering. The material in these guidelines is based on lectures which the Author was invited to give at City University, London, as part of the University's MSc course in Water Resources Engineering. The lectures commenced early in 1995 and continued yearly to 2000. The text was initially prepared in 1994 and has been updated and amended annually since then, which has resulted in numerous relatively minor though significant changes.

The knowledge offered in these guidelines is expected to be of practical rather than theoretical or academic value but it is the Author's strong contention that those involved in reservoir engineering should be able to understand one another and each other's tasks and responsibilities, whether they are engineers, administrators, specialists, economists or researchers.

Acknowledgements

The presentation of these guidelines has drawn on comments offered by experienced colleagues, particularly in specialist subjects, and on verbal and written material by those and other experts. Such contributions are gratefully acknowledged and particularly those of Professor K V Rao who co-operated throughout and contributed much of what is written on hydraulic studies in Chapter 5.

The Author is indebted to leading hydrologists Ian W Rose and Peter T Adamson for comments and advice on hydrological aspects, in particular on studies to determine design floods. The account of the contents of the Reservoirs Act 1975 in Chapter 17 is largely derived from the published work of J W Phillips. The following are thanked for their responses to requests to include material of which they are the originators or previous publishers.

- Centre for Ecology and Hydrology, Wallingford OX10 8BB, UK: Figs 6.4, 6.6, 6.7.
- Wilmington Publishing Ltd, Dartford DA2 7EF, UK: Data and descriptions of seismic incidents in Chapter 9 from *Dam Engineering*, February 1993, Dams and earthquakes — a review.
- John Wiley & Sons Limited, Chichester PO19 1UD, UK: Fig. 9.11, N N Ambraseys and J M Menu, *Earthquake Engineering & Structural Dynamics*, 1988, Reproduced with permission.
- International Commission on Large Dams, 75008 Paris, France: Figs 4.1 and 4.3, A geomorphological approach to the assessment of reservoir slope stability & sedimentation, *Proceedings of the Fourteenth Congress on Large Dams*, Volume III, Q54–R11.
- International Association of Hydrological Sciences, Ecole des Mines de Paris, 77305 Fontainebleau, France: Fig. 6.1, Copyright IAHS Press, 1989.
- Geotechnical Engineering Division, American Society of Civil Engineers, Washington DC 20005, USA: Fig. 9.37, Seismic design of concrete face rockfill dams, by Prof. H Bolton Seed *et al.*,

Symposium '*Concrete Face Rockfill Dams: Design Construction and Performance*,' ASCE, 1985.
- The Institution of Engineers, Sri Lanka: Figs. 11.5–11.12. *Journal of Institution of Engineers, Tunnelling Seminar*, Vol. VIII, No. 3, September 1985.
- Building Research Establishment, Garston, Watford WD2 7JR, UK: Figs 9.2–9.4, 9.6, 9.9, *An Engineering Guide to Seismic Risk to Dams in the United Kingdom*, 1991.
- Halcrow Water, Swindon SN4 0QD, UK: Figs 10.2–10.4, 11.1–11.4.
- Binnie Black & Veatch Ltd, Redhill RH1 1LQ, UK: Fig. 6.2, Binnie & Partners, 1980.
- Yorkshire Water Services Ltd, PO Box 500, Western House, Western Way, Halifax Road, Bradford BD6 2LZ, UK: Fig. 3.10.

The Author and Publishers have endeavoured to give proper acknowledgement to all the sources of material incorporated in these guidelines. They regret if any instances are found where they have inadvertently failed to do so and request that information of such cases be conveyed to them so that any necessary additions and corrections can be made during future editing.

Part I

Conceptual and planning practice for reservoirs

1. Introduction and philosophy of approach

Introduction
The objective of these guidelines is to convey subjectively the thinking on reservoir engineering practice developed and learned through long personal experience — experience involving people, civil engineering and related subjects. In reservoir engineering practice there is a need for individuals with a great variety of talents and preferred occupations and preoccupations but, however specialized these may become, broad and lateral thinking is something crucial that must also be cultivated. It is hoped that this book will assist readers to consider where they stand in reservoir engineering practice and in what direction (if any) they would like their involvement to progress. For those in already established positions, it is an opportunity to stand back and review their own approach and their relationships with others.

Background
Apart from a period in the Army as an electrical and mechanical engineer, the Author's career has been as a consulting civil engineer largely concerned with dams, reservoirs, hydroelectric power and irrigation projects but also at times with much involvement in and responsibility for specialisms such as surveying, finance and economics, geotechnical engineering, meteorology and hydrology and hydraulics, not to mention setting up and running libraries and the administration of people and departments. The implementation of a civil engineering project requires first of all identification of the need for it, most often by politicians or government authorities, followed by conceptual planning and prefeasibility studies, followed by detailed feasibility studies and, if the project is found to be wanted and economically feasible and there is sufficient finance in prospect, then design, approval by prospective owners and government authorities, tendering by contractors, followed by construction, commissioning and operation. It is only the

latter 'operation' which produces the material benefits from a completed project although benefits accrue on the way by means of work generation and employment and the development of technology during research and design.

Disciplines needed

It can be seen that, in reservoir engineering, there is broadly a need for civil engineers, specialists — including surveyors, geologists, geotechnical engineers, hydrologists, environmentalists, economists, technicians and inspectors — and researchers. In general, they may choose to work for specialist firms, research authorities, consultants, contractors, government departments or for private sector firms. According to their natural inclinations and talents and to the job opportunities available, civil engineers tend to work in the main fields of planning and investigations and/or in design, the preparation of contracts and management of contracts and/or in supervision of construction, or in the supporting academic field, including hydraulic and structural model testing, research, computer programming and numerical modelling. Depending on job opportunities, there are thus needs for competent, capable, professional people of whom desirably a relatively small number should prove to be geniuses and a large number will benefit from showing a flare for part of their work. The majority will, however, be workers because, if the work is there, it has to be done. Not all of it can be of immediate appeal and interest to those who must be asked to do it. It therefore also happens that to deal with a workload, specialists are called upon to deal with perhaps routine work outside their specialism and interests. It might be thought that the key to the success of a firm of consultants would be to have a core of brilliance, even of genius, because the business of consultants is to offer the services of people in engineering and to keep ahead of competing firms. However, those to be found at the top are not always the geniuses or the most brilliant but some of them have risen from the category of 'workers' — ones who have, perhaps, some gift for gaining the liking and respect of others and for successful management of work and people. There are thus opportunities in reservoir engineering practice for many kinds of men and women.

It used to be the case that employment with government authorities provided steady permanent work prospects (conducive to stable lifestyles for families) but was relatively rather stereotyped and unexcitingly paid, while more varied and interesting work, including work overseas, came to firms of consultants, and the hardest work and best remuneration, though short lived (depending on work load) would be with firms of contractors. To some extent this underlying situation still exists but all the possible forms of employment are more and more conditioned by commercial considerations. All employers try to find continuing employment for those whose services they most value but, due to the high cost of employment, there is inevitably an increasing incidence of hire-and-fire. One must be aware that those in civil engineering are less well remunerated than those in some other professions, except perhaps at the top. This is partly because of the competition that firms have to face (which, while bolstering the remuneration of the successful, diminishes what firms can competitively charge) and partly because of lack of direct contact with and appreciation by the public. Higher remuneration is obtained for work overseas but such work carries with it the cost and disadvantages of separation from one's home. In any case, firms have to pay the market rates to obtain the staff they need. In 1997, a survey reported the *average* earnings of Civil Engineers in the UK to be £35,600. Quite detailed information can be obtained from the *New Civil Engineer (NCE)*/Institution of Civil Engineers (ICE) salary surveys, previously published yearly, for example, 1999, Emap Construct Ltd, £150 for the full survey; a 16 page A5 summary issued with *NCE* March 1999).

Why then do civil engineers choose to work in the field of reservoir engineering? The Author believes that the answer is that water is essential to life and associated with that, it has an inherent appeal to the senses. Reservoirs tend naturally to be sited in beautiful surroundings: within hilly or mountainous terrain. Sometimes they are very remote from infrastructure and amenities (in the form of urban development) and hence there is an element of pioneering attached to their creation. There is an element, too, which must appeal to civil engineers, of harnessing the forces of nature to the use and convenience of mankind (the maxim of the ICE of the UK). Reservoir works involve elements of innumerable kinds of speciality — above ground, underground and underwater. There are challenges in minimizing costs of both the design and construction, of dealing with contractual issues, including settlement of contractors' claims (which requires something of the wisdom of Solomon) as well as a sound understanding of the issues and the contractual conditions by which they are governed. Water itself is a fascinating and challenging liquid — it is the medium of the hydrological cycle and a liquid about which an ever-increasing amount, but still too little, is known. It exhibits logical and random behaviour in phases.

Constraints on training and experience
While all kinds of work in reservoir engineering exist in the UK, and while smaller reservoirs generally present the same problems as

larger reservoirs, albeit with smaller resources to deal with them, the most challenging problems in reservoir development today exist in developing countries. There are economic, political and social reasons why developing countries would like to be able to engineer reservoir development in their own countries and, increasingly, with the development and accessibility of educational facilities and experience among their own nationals, they are more able to do so. Nevertheless, the preponderance of financial resources, experience and technical expertise remains with specialists, consultants and contractors based in developed countries. Certainly developing countries are reluctant to employ foreigners to carry out work that they believe could be undertaken by their own nationals (at much less expense). Foreign employers are under pressure to employ to the maximum and to work with nationals of the countries in which they are engaged. As a result, it is often decided that appropriate offices should be set up to have the work carried out in the country where the project is to be built. There is thus difficulty for young, inexperienced, persons (aged up to, say, 30–32) of western countries to gain the experience that they need in developing countries, unless their work is subsidized by their employers or by the development agencies of their own countries. Against this constraint is the principle generally followed by international financing agencies, that the engineers responsible for projects which the agencies are to finance, should not be nationals of the country seeking their financial assistance. An element of independent control of the utilization of the finance is thus secured, the original source of the finance being the various member countries of the agency concerned, which are anxious to be assured that the finance is put to the intended use. It can be seen that the outcome for engineers in developed countries is constraints on the opportunities for working in developing countries — opportunities which have to be justified by specialist qualifications or experience — or by greater efficiency in producing results than would be expected from local engineers. If these justifications are not demonstrable, it would be argued that money is being mis-spent.

The Engineering thread

The Engineering thread, with a capital 'E', is the desirable continuity of engineering knowledge and thinking which should exist throughout the stages of planning, design, supervision of construction and operation of a project. The need for it is to ensure that no important assumption or precept is overlooked at any stage because this might lead to design, construction or operation which is not soundly based and could lead to a wrong or even an unsafe outcome. The thread can be woven partly with the essential aid of the preparation of comprehensive records and reports prepared at the time, while the work is in progress and particularly on completion of each stage. It cannot, however, be positively ensured that every important detail is clearly recorded or even if it is, that some person expected to implement some item during a later stage has read and understood the necessary detail. There is thus a need for continuity of involvement of the individual(s) responsible for planning and designing the project. It was, at one time, wrongly believed by financing agencies and, perhaps even now, by some governments, that the same firm of engineers should not be responsible for successive stages of the works, for example the same firm should not be employed on both feasibility studies and design. The reason was partly to give other firms the opportunity to share in the work and experience but also especially to introduce new thinking and in this way to obtain a second opinion. There is indeed a risk that the involvement of only a single engineer or small group may result in some important consideration being overlooked due to lack of some particular experience or due to prejudice or a closed mind. This risk can perhaps best be dealt with (and now often is) by the appointment of small panels of independent experts who assemble every three to six months or so to review the work which has been carried out since their previous assembly, to discuss it with the Engineer and Employer and to report to the Employer and to the Engineer on their findings. This procedure should in no way relieve the Engineer or the Employer of their responsibilities and thus cannot be binding on either party but clearly the panel's recommendations and advice must be fully considered and will only rarely be disregarded unless for justifiable reasons. The independent experts should be selected because of their special ability and experience in the work being undertaken. Their numbers need not usually be more than three or four and there should be continuity of membership of the panel throughout the work to be implemented. Although it may not be appropriate for the same members of the panel to assemble each time, it is desirable that the same leader of the panel should always be present (to maintain the 'thread' and cohesion). The leader is usually a 'generalist' with wide experience, capable of taking an overall view. Other members should be specialists in the critical work that has been undertaken or is being undertaken at the time. It is also important, for the progress of the work, that the panel be able to assemble at appropriate critical times. This is sometimes difficult but it is important because, if the panel discussion results in the recommendation of changes, there will be delays and unnecessary costs if recommendations arise too late.

Mode of work

The different kinds of specialisms and disciplines that are required in reservoir practice have been outlined and the varying nature of the different kinds of work that have to be undertaken has been indicated. Less than 40 years ago it was possible for engineers to be 'generalists' and to achieve a leading place in their profession, having mastered a range of the governing subjects. In some cases, a feasibility study for a reservoir project could be undertaken on the basis of a visit to the site by only one engineer, whereas currently it is prudent for the site to be visited by a 'generalist' engineer, supported by a hydrologist, geologist and economist to say the least. Calculations could be carried out by using slide rules and log tables or even entirely manually. By wading through paper, it was usually possible to appraise visually and check the calculations and to assess the basic principles and assumptions on which they were based. From what has been said so far, it will hopefully be seen that there was an advantage in having the wide grasp and understanding of the whole subject which it was possible for an engineer to obtain. With, however, the development of knowledge and experience in more recent years, the amount of information which has become necessary to comprehend specialized areas of reservoir engineering has resulted in very few engineers (if any) having sufficient *overall* grasp of these increasingly specialized subjects.

Fortunately, as is so often the case and the way with nature, resources have been discovered and developed to fulfil a need: the service of computers and use of the internet has accelerated almost beyond bounds the work that can be undertaken and the problems that can be solved within available time limits. These have also made possible more rigorous and satisfying methods of analysing, processing and solving problems. It can, however, be seen that, as well as placing more tools readily in the hands of individuals, they have accelerated the potential for the growth and complexity of specialized subjects. In most cases, individuals must use programs developed by others, simply because there is not the time, need, or often the ability to recreate programs from scratch. It should, however, be an accepted principle that those who use a program and data must be satisfied that the principles and assumptions on which they are based are suitable for their application, and that the programming for that application has been validated by successful use on previous similar applications. It should also be an accepted principle that the output in each case is subjected to an independent check, coupled with an appraisal to confirm that the results appear sensible. In one memorable case, the output (from a university) of the finite element analysis of an arch dam showed tensions where one would expect to find compressions. It was only belatedly in a final appraisal that any comment was made on this, as a result of which it was found that the program had applied the reservoir water load to the downstream face of the dam — in an upstream direction...

The need for mutual understanding

What has just been said is intended to convey the need, but increasing difficulty, for engineers and specialists to understand each other's language, mode of working and objectives. Specialists must comprehend the significance and importance of what they are doing for the project without spending time on aspects that have little practical effect on the outcome. They need to understand the effects and implications of the work which they do themselves, on the outcome of the work of others, and they need to appreciate the effects and implications of the work that is done by others on the work that they have to do themselves. There must therefore be a continual process of interchange, adjustment and avoidance of abortive work. The process involves reading and understanding the relevant essentials of reports by other members of the team who are working on the project. Weekly meetings between the key members of a team may be a desirable feature in this respect. In order not to get lost, it is desirable to acquire a basic knowledge of the terminology used by other specialists and generalists.

The progress and direction of the work throughout is commonly much influenced by economists and politicians. Even the technical specialists on the team need to be able to talk to such people and to understand what they are saying and thinking and why. In turn, the technical specialists need to make themselves understood, so that their work can be properly taken into account. There was a case, for example, when a brilliant young hydrologist and numerical modeller went to a faraway developing country to carry out an analysis of the hydrology for the development of hydropower on a river. The results were not accepted, not because they were technically unsound (they were sound) but because they were not understood to the Employer's representatives and those representatives had not been convinced that the hydrologist had understood and taken into account factors that they thought were important. The work had to be done again with imaginable delay and increase in cost. A likely outcome in such a case is that the person responsible for the rejected work becomes persona non-grata for further work on the project. It is fortunately remarkably often the case that the most talented of academics and specialists have the gift of making themselves readily understood to people who are quite unfamiliar with their subjects. There is certainly merit in including in reports and technical papers a glossary of terms likely to be unfamiliar to even only a few readers and a list of the words denoted by abbreviations. Readers can be irritated and distracted by having to search elsewhere for such explanations. It is as well to bear in mind that perhaps the most important readers are decision makers who very often have not had a technical education and have limited time to spend on reading reports. This points to the importance of incorporating a concise summary of a report at the beginning with a list of conclusions, taking up perhaps only one, two or three pages. It is, of course, often the case that one species can only talk to, understand and be convinced by, others of their own species. A prime example of this is the economist species. It is, nevertheless, understandable and natural that an economist or other specialist should have respect for and find easily intelligible the views and findings of others qualified in the same specialism. Even though often they may not agree with one another.

The roles of empirical and theoretical methods

For most professional engineers and scientists, only theoretical solutions are satisfying because they provide an understanding of the problem in question and an assurance that the solution is logical. However:

- it is necessary that all theoretical solutions are validated in practice
- there are not theoretical solutions to all problems
- physical modelling gives the satisfaction of providing a visual demonstration of a problem and its possible solution(s).

It is therefore inevitable that there must be conjunctive use of both theoretical and empirical methods and in some cases it is necessary to resort solely to empirical solutions. Sometimes there is a choice but one does not necessarily exclude the other. This can be illustrated by the following two examples.

1. Mathematical models and hydraulic survey data were used conjunctively to model hydrological conditions in the Irrawaddy River Delta in Burma (now Myanmar). The delta extends over 31 000 km^2 and comprises four linked sub-deltas with fluvial–tidal interactions; 12 major mouths extend over 260 km of coast (Fig. 1.1). The shaded areas represent higher ground outside the delta. The object was to set up numerical models capable of predicting the effects of engineering works (including the possible construction of embankments for flood control and channels for

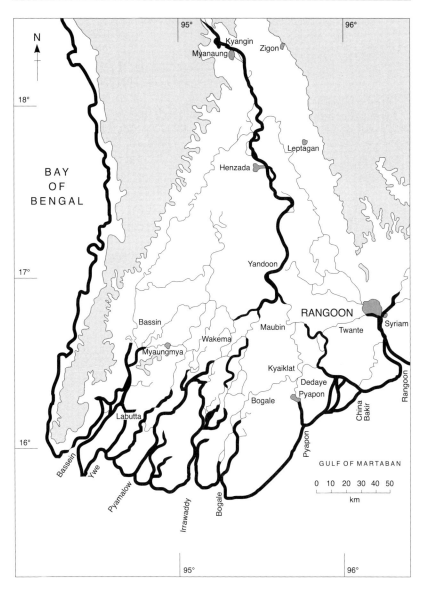

Fig. 1.1. Location map: the Irrawaddy Delta, Myanmar

flood relief), to predict the effects on flood levels during both the floods and their recession and on tidal flows and saline intrusion in the delta, and hence to plan the future development of the delta area for reclamation, irrigation and other forms of development. There are some 3200 km of navigable channels in the delta, so a considerable task was the acquisition of topographic and hydraulic survey and other data, including tidal variation data. River cross sections were surveyed at 663 locations at an average of 4 km intervals. The simplification and minimization of such requirements as far as possible was obviously an important consideration. The mathematical models were, of course, based on established hydraulic principles but the calibration and final validation of the models by empirical observations were an essential part of the process.

Five mathematical models were developed for different aspects of the study:
(a) a pilot model, using existing data, old charts and satellite photography
(b) a dry-season model with improvements from field observations
(c) a surge model simulating surges generated by a cyclone in the Bay of Bengal
(d) a saline intrusion model to establish the seasonal movement of salt
(e) a wet-season model representing river cross section properties at high flow levels.

The work has been described in full in the literature [1].

2. Horizontal cracks unexpectedly appeared and extended in a mass concrete gravity dam in Scotland, allowing leakage from the reservoir. Investigations could not determine the cause of the cracking but the observed behaviour of the dam was successfully reproduced by increasingly sophisticated mathematical modelling. For this it was postulated that there were potential geophysical inward and outward movements of the valley sides (of which there was survey evidence) coupled with movements due to seasonal temperature changes. The key to the phenomena was the observed closure of previously open contraction joints in the dam due to the deposition of calcium carbonates, which solidified in the joints, preventing longitudinal expansion movements in the dam. Finite element modelling of these conditions convincingly reproduced the behaviour of the dam with regard to cracking and stress development and provided the basis for design of remedial measures and the prediction of future behaviour. The procedure was, in effect, a 'back-analysis'. The problem and its solution were quite fully described in *Floods and Reservoir Safety* [2].

So, how much does an Engineer need to know? At the highest level, there needs to be an awareness of the best international practice and thinking. This entails being informed of the output of relevant publications and conferences — a major undertaking, considering the number of these. Even relatively junior staff may be called upon to assist in acquiring and disseminating such knowledge. The internet provides an enormous source of knowledge and data but identification and sifting of relevant data forms a formidable undertaking. To be unaware of important developments can lead to failures in engineering practice and to the most serious allegations of negligence in keeping up with good practice.

Firms taking the time and trouble to obtain certification by the UK Accreditation Service (UKAS) to ISO 9000, give their employers the assurance of independent approval of their quality control procedures to a high standard.

The conclusion on this theme is that the optimum use should be made of empirical and theoretical methods and mathematical modelling and that they should either be used conjunctively or considered as alternatives.

References

[1] Brichieri-Colombi, J. S. A. (1983). IAHS Publication No. 140, *Hamburg Symposium*, August 1983, 353–364.
[2] Gosschalk, E. M. *et al.* (1991). Overcoming the build-up of stresses, cracking and leakage in Mullardoch Dam, Scotland. *17th ICOLD Congress, Vienna*. Q65, R-26, 475–498.

2. Objectives

The purposes of reservoirs

The fundamental purpose of a reservoir is to store water and the size of the reservoir is governed by the volume of water that must be stored, which in turn is affected by the variability of the inflow available for the reservoir. Reservoirs are of two different main categories: (a) impounding reservoirs into which a river or rivers flow naturally and (b) service or balancing reservoirs receiving supplies that are pumped or channelled into them artificially. In general, service or balancing reservoirs are required to balance supply with demand, in a system where there are long conduits through which, without them, response to demand would be excessively slow. Such reservoirs are usually relatively small because storage for a matter of only hours or a few days is sufficient for their purpose. Impounding reservoirs have the purpose of regulating the flow in rivers, whether to meet demand for water or to attenuate and control the flow during floods. The storage capacity in a reservoir is notionally divided into three or four parts (Fig. 2.1).

- *Live storage* — the storage available for the intended purpose between full supply level and invert level of the lowest discharge outlet. The full supply level is normally that level above which over-spill to waste would take place. It is sometimes called 'normal retention level' or 'top water level'. The minimum operating level must be sufficiently above the lowest discharge outlet to avoid vortex formation and air entrainment.
- *Dead storage* — the total storage below the invert level of the lowest discharge outlet from the reservoir. It may be available to contain sedimentation, providing the sediment does not adversely affect the lowest discharge outlet.
- *Flood storage* — which may be needed in reserve between full supply level and maximum water level to contain the peaks of floods that might occur when there is insufficient storage capacity for them below full supply level.

16 | Reservoir engineering

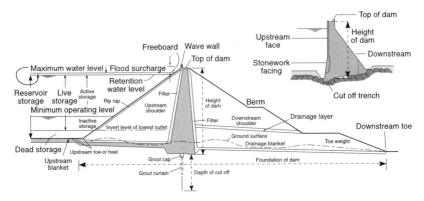

Fig. 2.1. Dam terminology

- *Freeboard* — the margin for safety between the level at which the dam would be overtopped and maximum still water level. This is required to allow for settlement of the dam, for wave run up above still water level and for unforeseen rises in water level, because surges resulting from landslides, earthquakes or unforeseen floods or operational deficiencies, etc. In some cases, notably in the 1996 ICE publication *Floods and Reservoir Safety: 3rd Edition* [1], the dam's freeboard has to contain flood surcharge but this seems misleading with regard to the limits of land that are liable to be inundated by the reservoir. ('Normal retention level' is sometimes called 'top water level' but this is misleading because the 'normal level' can be exceeded up to maximum water level during floods.)

The functions of reservoirs are to provide water for one or more of the following purposes. Reservoirs that provide water for a combination of these purposes are termed 'multi-purpose' reservoirs.

- *Human consumption and/or industrial use.*
- *Irrigation* — usually to supplement insufficient rainfall.
- *Hydropower* — to generate power and energy whenever water is available (in which case, alternative supplies of power and energy are required when water is not available) or to provide reliable supplies of power and energy at all times when needed to meet demand (in which case sufficient storage is necessary to provide water for generation during periods of demand when otherwise there would be a shortage of water).
- *Pumped storage hydropower schemes* — in which water flows from an upper to a lower reservoir, generating power and energy at times of high demand through turbines, which may be reversible,

and the water is pumped back to the upper reservoir when surplus energy supply is available. The cycle is usually daily or twice daily to meet peak demands. Inflow would not be essential were it not needed to replace water losses through leakage and evaporation or to generate additional electricity. In such facilities, the power station, the conduits and either or both of the reservoirs could be constructed underground if it was found economic to do so.
- *Flood control* — storage capacity is required to be maintained to absorb foreseeable flood inflows to the reservoir, insofar as they would cause excess of acceptable discharges through spillway openings. This storage also allows future use of the flood water retained.
- *Amenity use* — such as boating, water sports, fishing, sight seeing, etc. Some of these activities may be developed commercially.

It can be seen that some of these objectives may be incompatible in combination. For example, water has to be released for irrigation to suit crop growing seasons and shortage of rainfall, especially during droughts, which may occur for short periods or may prevail over several years, while water releases for hydropower are required to suit the time of public and industrial demands. The latter will be affected not only by variations in economic conditions but also by variations in the hours of darkness and air temperatures — demand for power for air conditioning increases in hot weather and demand for heating increases in cold weather. Recreational use and tourism peak during fine weather and holiday seasons. Frequent draw down of reservoir levels for water supply, irrigation or hydropower can have effects on the shore line that are adverse to amenity use. Thus, in planning a reservoir, the engineer must be clear from the start what objectives the reservoir is required to serve. These have usually been decided by the Engineer's employer, 'the undertakers' who are usually the owner(s) and/or operator(s) who has found or been advised that a reservoir could be a means of fulfilling a need or promoting a purpose. It is necessary to be clear whether the project is to serve local needs or to contribute to serving regional or national needs, or whether the output is to be exported to earn foreign currency. Very often the undertakers will have had terms of reference prepared to describe in some detail what engineering or other services they are seeking.

The engineering planning of reservoirs

The first step in planning is therefore discussions with the undertakers to make sure that their needs and purposes are fully understood, together with known constraints (which may be financial), the available time until interim reports and the final benefits are known to be needed, and social, religious and political constraints. In many countries, including the UK, there are areas which are unsafe to visit, much less work in, without suitable precautions against opposition. The modes of travel and transport may be restricted. The availability and feasibility of using helicopters in the area of the project may be important. There may or may not be a bias towards employing local people and contractors, whose abilities and resources therefore need to be known. Reduction in unemployment and minimizing imports may be important objectives of the employer, as may an increase in foreign tourism. Completion may be desired before the time of a political election or before some significant high-level meeting. All these factors may affect the speed at which it is possible to progress the works as well as the nature of the works themselves. Labour-intensive types of design and forms of construction using indigenous materials may or may not be indicated. Some or all of this information should be obtainable from the terms of reference if they exist.

The second step is the assembly of all relevant existing information:

- reports, if any exist, on previous investigations and studies, including a master plan for the region and/or river basins of interest
- reports on projects similar to that proposed which have already been constructed in the region or which are planned
- a computerized geographical information system for areas of interest, if such exists, which may contain much of the information required
- topographical data in the form of maps and aerial photographs (in stereo pairs if available)
- geological data in the form of maps and the results of drilling and testing in the project area
- data on historic seismic activity in the region
- meteorological and hydrological data — what records are available of parameters such as rainfall, atmospheric and water temperatures, evaporation, humidity, wind speeds, hours of sunshine, river flows and/or river levels, and sediment transport, and over what periods are the records of each parameter available?
- data on alternative supplies, including imports, which could supplement or replace outputs from the proposed project
- for water supply projects, data on population and forecast population growth, industrial water requirements and forecast industrial developments
- for irrigation projects, data on soils in the project area and on crops already grown, including water requirements for those crops

- for hydropower projects, data on past demand and forecasts of future public and industrial demand for power and energy; data on existing transmission systems, including transmission voltage and capacity
- data on flora and fauna in the project area and on fish in the rivers and lakes, including data on their migratory and breeding habits
- data on tourism and recreational use of rivers and lakes and if and how this is to be encouraged to develop in future.

While quality assurance has improved significantly in recent years, especially in developed countries, in the past pains have had to be taken to bring to light serious errors in data of all kinds and in reports by even highly reputable engineers from developed countries. The subject of the need for data validation and improvement will arise again later, particularly in Chapter 3.

Plan of work

As a result of the work previously described, all the available information should be to hand and the engineer can appraise the existing site and begin to conceive the works, including alternatives, which may be required and which seem potentially feasible. It may at this stage be necessary to formulate proposals and to put these to the employer, with programmes and estimates of the cost of undertaking the work of investigations, prefeasibility studies, feasibility studies and/or preparation of tender documents and design. These proposals may have to be made in competition with other firms with a resulting challenge, in making the proposals, to ensure and demonstrate that the work will be sound, of satisfactory quality and in accordance with the best practice, while keeping the cost to a realistic minimum. Employers may be pleased by assurance that they will be kept regularly informed and consulted at meetings and by the submission of reports. Innovations that have technical advantages and would show savings in cost may earn credit compared with competitors but the employer will wish to be reassured by past successful experience in the kinds of work proposed and by the experience of the staff to be engaged. It is worth restating here the well-established principle that skimping on quality and quality assurance to save costs is a false economy, because errors cannot always be corrected without serious consequences, not to mention the loss of time and money which may result, in the short or long term. In this connection, the engineer must think laterally and broadly to be able to foresee, as far as possible, potential adverse effects and hence to be able to contend with them. These adverse effects might entail adverse weather conditions (when certain types of work would be impracticable on site), shortage of labour and/or of certain skills, delays in obtaining approvals of the employer to place orders for surveys and site investigations, difficulties in obtaining approval for procedures to use explosives, difficulties due to civil disturbances and so on. One must, however, avoid becoming obsessed with very many numerous potential difficulties and remember that, while some difficulties are usually inevitable in reservoir practice, the risk of each additional difficulty happening as well becomes factorially smaller. If one became overwhelmed by all possible difficulties, nothing would ever get done. The objective is to eliminate or to be prepared to deal with as many as possible of those difficulties which can be foreseen. An overall risk analysis should show whether foreseeable combinations of risks are acceptable — but an exhaustive analysis can be exhausting.

Programming the work of reservoir engineering can be considered within the following time scale: since use of time incurs expenditure on non-productive costs and there is a financing cost on the money committed, it is usually economical to expedite the work even though this may entail the use of increased manpower and equipment. However, this does depend on being able to use the manpower and equipment efficiently and to avoid, as far as practicable, their employment on work that proves abortive because sufficient time has not been allowed to ensure that the work was justified.

Time scale
1. Prefeasibility study — without any site investigations except auger holes and shallow trial pits and simple testing and without work on designs other than from experience or analogy with other projects: may take   1–3 months
2. Approval to proceed with next stage: may take another
                                                              1–3 months
3. Feasibility study — depending on requirements for surveys, site investigations, additional data required, conceptual analyses and feasibility design: may take   4–12 months
4. Approval to proceed with next stage   3–6 months
5. Detailed design and preparation of tender documents
                                                              6–12 months
6. Approval to proceed with tendering (which normally implies the procurement of finance for construction)   3–6 months
7. Tendering and award of contracts   3–6 months
8. Construction and commissioning (including the time required for the reservoir(s) to be filled at least to minimum operating level(s))   12–48 months

It can be seen that the overall time required for implementation of a reservoir project can be expected to lie within the limits of 33 to 96 months or say, three to eight years. Clearly, the lower end of the scale applies to a relatively small straightforward project while the upper end applies to large schemes. No allowance has been included for unnecessary delays or exceptional difficulties during construction. To restrict the commitment for financing by eliminating or reducing work on unacceptably costly alternatives, additional stages for basic design and its approval, say 6–8 months overall, are sometimes called for before stage 5 (detailed design) but this should reduce the upper limit of time required for stage 5 by, say, three or four months.

It should be noted that it may be necessary to hold a public inquiry and/or to obtain statutory authority before stage 7 (tendering) or at least before any award of contracts for construction. On the other hand, if finance is available and time is short, stages 1 and 2 may be omitted by incorporating them in stages 3 and 4. It follows from the time scale that preliminary work for a reservoir project has to be started three to eight years before the outputs are needed and that the size and capacity of the projects has to be sufficient to meet the growth in demand for their outputs which is likely to occur before an extension of the project or other contributing projects can be completed. The size and capacity of the projects are likely to need to include a margin of capacity to cover the output of part of the projects or of other contributing projects in a system, when that part is out of commission for maintenance or repair or for other reasons. This margin might notionally be taken as 10% of the output of the system or, more reasonably, the output of the largest section of the system that might be put out of use at one time.

The work to be included in each stage will be covered in more detail in Chapter 3.

Reference

[1] Working Party chaired by Carlyle, W. S. (1990–1994) followed by Kennard, M. F. (1996). i–viii & 1–63 on Floods & Reservoir Safety (1996). *Floods and Reservoir Safety*. Thomas Telford Publications for the Institution of Civil Engineers, London, 3rd Edition.

3. Selection of potential dam sites and conceptual schemes

Appraisal of data and formulation of possible projects

In Chapter 2, the range of information needed for the formulation and design of reservoir projects was summarized. Once the information is assembled and the objectives are clear, appraisal of the information will usually serve to identify the most promising possible projects and alternatives. Even if sites have been proposed in previous reports, the proposals should always be reviewed and looked at critically and the information published should be spot checked and progressively checked, depending on whether or not errors are found and on their scope.

If there is a computerized geographical information system for the areas of interest, its contents should be identified. It could indeed contain a great deal of the information required, including the key characteristics of potential dam and reservoir sites of interest, such as maximum heights of dams and maximum volumes of reservoir storage, at least where governed by the topography. A master plan might be accompanied by a numerical resource allocation optimization model (RAOM) of the proposed development by which a combination of possibilities can be optimized. The sufficiency and soundness of such a model should be considered. For reservoir sites, there can often be conflicting criteria, as described below.

- For the greatest water availability, one is looking for the maximum possible catchment area, which implies sites at the lower end of a river basin.
- For irrigation projects, it is desirable that the minimum operating level of the reservoir should be sufficiently above the irrigable area so that water can flow by gravity, allowing for head losses in transit, to the furthest reaches of the area, including spreading over the fields, without the need and expense of pumping.

- For hydropower schemes, in simplistic terms one is looking for the maximum potential average power output, which is proportional to the product of average river flow and the potential fall in the river, or 'head' over which the power can be developed. The maximum head would usually be available from dam sites at the upper end of the river basin where the catchment area, however, is relatively small. The optimum combination of river flow and head for hydropower schemes is thus usually found at the lower end of a range of hills just above the lower basin or flood plain.
- For dam sites one is looking for a neck in the valley sides where spurs of the hills encroach close to each side of the river. This offers a short length of dam and suggests resistance to erosion and possibly a satisfactory foundation for the dam on its flanks, although the river may have followed a course taking advantage of a line of weakness in the terrain and eroded its path through the spurs. This will need investigation.
- Possible dam sites need to be considered in relation to available geological information.

Foundations on sound rock are likely to be the most suitable preliminary choice from the points of view of strength and impermeability but some rocks are suspect. The most prevalent suspect rock type is limestone because it can go into solution in flowing water, resulting in so-called 'karstic' conditions with caverns and voids. These may be interconnecting, which would allow leakage of water under pressure, difficult to prevent by grouting. It can sometimes be very expensive or even impossible to make such conditions watertight. However, sound, fresh limestone can provide a good, watertight foundation, if it is not subject to flowing water; it can also be used as aggregate for making concrete. There are other rocks, such as gypsum, anhydrite and halite, which are also soluble. At this stage, the aim will be to look for encouraging or warning signs in the geological information. Site-specific studies will be needed and are discussed in Chapter 4 on geological studies. However, plans showing identified faults and lineaments, and stereo aerial photography and satellite imagery from which faults and lineaments may be identified, are valuable at the earliest stage because they indicate discontinuities in rock conditions and possible paths for leakage from a reservoir. It is possible that these features are still subject to tectonic movement or may be reactivated by earthquakes or even by seismic activity — induced by reservoir filling. Thus dam sites intersected by faults, or in the close vicinity of faults, are to be avoided if possible. Faults in the close vicinity of dam sites will need special studies and investigation.

Appraisal of data involves, as well as implications for the objectives of the project, the appraisal of the sufficiency of the data and their reliability. From this will arise determination of the need to procure additional or improved data. The first and simplest question is who was responsible for procurement of the existing data? The answer, however, will not be conclusive because, as already mentioned, even the most reputable firms and experienced individuals can and do make damaging mistakes and omissions. The next question is whether it is known if the data have been checked and if so, in what way and by whom? Different types of data will require different treatments, as outlined below.

- *Topographical data* — what is the coverage, what are the scales and what are the contour intervals? 1:100 is the smallest scale for detailed design of structures, while 1:1000 is suitable for longitudinal sections of conduits and transverse sections of dam sites but 1:20 000 maps or even smaller should be suitable for conceptual layouts of the projects. The contour interval usually represents the standard error in contour level. Thus, even for conceptual schemes, it is desirable for the contour interval to be less than about 10% of the potential heads to be developed or of the heights of dam under consideration. How accurately the overall survey datum has been established needs to be determined and to what datum it has been related. This is of less importance if all the projects of interest are satisfactorily related to a project survey datum and if they do not affect and are not affected by controlling levels outside the project area. On the other hand, a reservoir discharge might be affected by the backwater levels from a reservoir on the river below it, or its own backwater levels might affect the scheme above it. If additional mapping is required, aerial surveying usually presents the fastest means of obtaining it, although reliable ground control for aerial survey is time consuming if it does not exist. Aerial photography is also expensive, and dependent on satisfactory weather conditions for both flying and photography. The accuracy of ground levels given by aerial survey is affected by tree cover and by how carefully allowance has been made for the height of trees.

  For detailed design of an irrigation area, 1:5000 scale mapping (preferably orthophoto mapping) with 0·5 m contour intervals or less (even down to 0·1 m) is desirable, if the area is fairly flat. For feasibility studies, 1:20 000 with 1·0 or 0·5 m contours, depending on the nature of the topography, should serve. The 1:5000 mapping might be produced from 1:20 000, or larger scale, photography. Creation of a digital ground model could have considerable advantages and can be considered.

- *Geological data* — this is best reviewed by an experienced engineering geologist and discussed with the chief civil engineer. The need for additional geological survey will depend on the existing coverage and the extent to which geological features and detail are shown. Unforeseen geological conditions are a disturbingly common cause of additional costs and delays in the design and construction of dams. One of the most important considerations early on is the review of data and ground reconnaissance by the experienced engineering geologist who should understand geomorphology in order to be aware of the land movements which have created the topography and the weaknesses which they may imply. The geologist should have a reputation for his or her acumen. For dam design, there must be every effort to identify weaknesses in shear resistance and possible paths which could allow the development of leakage. These problems can be associated with foliation joints or shear planes, which may have clayey layers or clayey infilling, or they may be associated with open joints in the rock. Sliding, i.e. shear failure, is one of the most frequent mechanisms of failure in dams. Stereo-viewing of aerial photographs is an excellent means of identifying whether landslides or rockslides have occurred in the past, which at once gives a measure of the stability of slopes and an indication of ground conditions.

    Confirmation of the stability or otherwise of the slopes surrounding reservoirs will become important in later studies because slides of enormous volume can occur, reaching extremely high velocities, so causing tidal waves if the slides descend into a reservoir. The most notorious case of this was Vaiont Dam in northern Italy (Fig. 3.1), an arch dam 265 m high, which was submerged under a 100 m high surge wave due to such a rockslide. The dam was completed in 1961. A slide had occurred in November 1960 in which about 690 000 m^3 of mountain side moved, after which the allowable lake level was reduced. However, the limited level was exceeded after heavy rain in September 1963, by which time creep of the mountain side had reached about 10 mm per day. The reservoir level was lowered again but on 9 October disaster occurred. In the space of about 5 minutes, 240 million cubic metres of rock slid at velocities as high as 30 m/s. The dam did not fail, but the reservoir was divided in two by the slide. A total of 2600 lives were lost and the reservoir has been abandoned. Figure 3.1 also shows Mauvoisin Dam, which failed catastrophically in the south of France in the 1960s due to shear failure in the foundations.
- *Data on historic seismic activity* — if not available locally, this is obtainable for worldwide situations from the British Geological

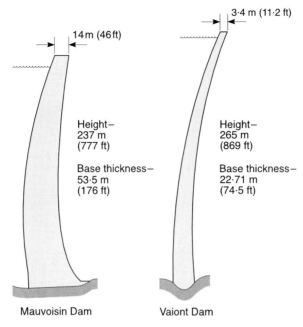

Fig. 3.1. Mauvoisin and Vaiont Dams: cross sections

Survey, Edinburgh; the International Seismological Centre, Newbury and other such authorities, which will also offer to analyse the data, if required. If there are not accelerographs in the area, for dams over 30 m high, installation of equipment should be considered well before the time comes for reservoir filling, and this can best be justified and decided at the start of the design stage. Guidance is given in *An Engineering Guide to Seismic Risk to Dams in the UK* [1].

- *Hydrological data* — this is best reviewed by an experienced hydrologist and discussed with the chief civil engineer and with whoever may carry out numerical modelling (if it is not the hydrologist personally). The objectives will be mainly to arrive at long average flows in the rivers of interest, potential low flows during periods of drought and extreme high flows during floods. The first stage is to review the existing records and to check and correct or omit erratic records which seem unreasonable — for instance if they are inconsistent with rainfall at the time. Gaps in the records have then to be infilled. Most simply, infilling can be done by interpolation, or insertion of average flows for the time of year but more convincingly it can be done by correlation with relevant rainfall records or with flows in neighbouring rivers. The next question is whether the records are over a long enough period to represent the likely range and average of future flows. This is a controversial question because various cyclical influences on river flows have been postulated and sudden changes in average flows have been observed, even in the flows of the Nile. Cyclical influences postulated have included:

- changes in the energy emitted by the Sun as measured by the rate of change of 'sunspot numbers': a cycle with a period of 5·5 years
- occurrence of the Humboldt current in the Pacific ocean: (El Niño) about every 9 years
- north and southward movement of an anti-cyclonal belt across the equator: about every 19 years
- climatic cycles represented by rings of growth in the trunks of trees: a period of about 35 years

The hydrological data should also include data on existing lakes in the area of interest — their storage capacity, fluctuations in levels, surface areas, open water evaporation records, sedimentation data, flood levels and wave records, if available.

There is also the not yet adequately known possibility of long-term climatic change. The economic life of a reservoir could be as little as 20 years but there may not be any conception as to how the reservoir could then be replaced. It is commonly considered that 80 to 100 years is an acceptable project life but some reservoirs have survived for more than 1000 years. From the above, it would seem that 35–40 years is a satisfactory length of river flow records, covering a range of apparently observed worldwide cyclical changes. The period over which rainfall records have been recorded is usually much longer than that of river flows and by correlation with the rainfall records, it should be possible to confirm if the period of river flows includes extremely wet or dry periods. The probable (or possible) maximum and minimum values cannot be expected to be found in a period of records and the problems of forecasting rare events are dealt with in Chapter 6.

As a result of the review it should be possible, if necessary, to improve the procedures for obtaining hydrological data and to change equipment and to extend the parameters observed. Time may be too short, however, before design work commences, to be able to depend on obtaining a representative period of observations. This also will be discussed in Chapter 6. There is no doubt that sufficient satisfactory observations should be put in hand for the future, if not for present design, at least for confirmation of assumptions and plans of operation.

At this point, it should be possible to plot longitudinal profiles of the river(s) of interest, preferably from existing mapping with contour

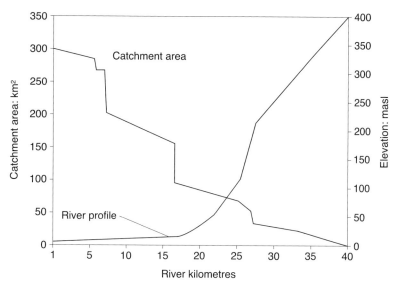

Fig. 3.2. Catchment area — elevation curve

intervals of 10 m or closer. The catchment areas at possible dam sites and at changes in topography should also be planimetered and superimposed on the profile(s) (as shown on Fig. 3.2). Dam sites will tend to be located below rather than above the confluence of tributaries so that the catchment of the tributary will be incorporated without the need to divert its flow or without otherwise losing it from the project.

Prefeasibility studies

'Prefeasibility' studies have the purpose of taking the studies of a possible project just far enough to determine if the undertaking of more detailed and expensive feasibility studies is justified. They are expected to give an indication of the likely technical and economic feasibility of the project but not to guarantee it. Usually they do not include site investigations or detailed design, but produce recommendations and perhaps terms of reference for the investigations and further work which would be required for detailed feasibility studies. They should include the preparation of preliminary layouts, a programme for construction, cost estimates and estimates of benefits. Uncertainties and possible difficulties should be discussed.

To some extent, the appraisal of data (just described) may need to be included as part of the prefeasibility studies (or conceptual studies) if it has not been carried out beforehand. This will depend on the terms of engagement of the Engineer. Once the appraisal and identification of potential reservoir sites has been carried out, the next step should be reconnaissance by a small team. This might best first be achieved by helicopter, homing in on sites, which can be examined on the ground after the helicopter lands. Depending on transport facilities, alternative modes of travel to the site by mule, foot, vehicle and boat, etc. are all possibilities in use. In somewhat unexplored country, video photography from a helicopter can be very helpful — for example when travelling up river to obtain a continuous scan of the changing topography.

The reconnaissance 'team' might possibly be just one person — the chief experienced reservoir or project civil engineer. Ideally, however, the smallest team desirable would include the chief engineer accompanied by the experienced engineering geologist and the hydrologist. The appraisal of data and possible sites will have alerted them to what they need to be looking for and anything unexpected or inconsistent with the data should be questioned and considered. Supplementary auger boreholes may be made and shallow trial pits may be dug. The team must look for surface evidence and take small samples of soils and rocks for reference and perhaps for particle size analyses, index tests, unconfined compression tests and so on back at base. Rock and soils will be exposed during road and other excavations, surface exposures of rock, joints and faults may be seen (take a sample of any clayey infilling (gouge)), scars of landslides may be seen, evidence of flood levels may be noted by flotsam left suspended in trees or high and dry elsewhere, sometimes notices have been erected to record notorious historic high flood levels — perceptive detective work is needed. The team should talk not only to local authorities but to local residents to probe their memories and knowledge about rainfall, droughts and floods, winds, earthquakes, etc.

Inquiries should be made about local resources and materials, including about local contractors who could help with investigations and construction.

It may be that the reconnaissance will cause certain sites under consideration to be rejected because topographical or geological conditions would render them excessively expensive, and uneconomic to develop. If so, it will be fulfilling its purpose by avoiding further abortive work and expense on those sites. On return to base, the team should compile a report on their reconnaissance.

If there is one main basis of reservoir development, it is hydrology. Revealing and key tools for the assessment of hydrology for reservoirs are flow duration curves and mass curves prepared for the river or rivers of interest at the possible dam site(s). These should be prepared at a very early stage on the basis of the historical river flow records.

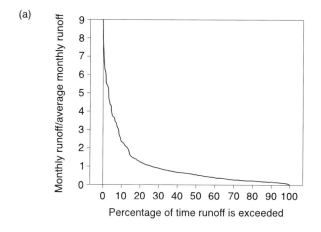

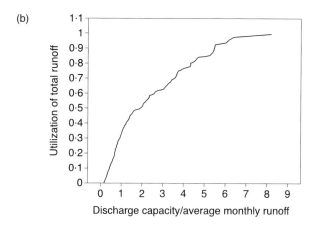

Fig. 3.3. (a) Unitized monthly flow duration curve. (b) Unitized monthly utilization curve

They can be revised and improved later as the development of data permits.

- *Flow duration curves (e.g. Fig. 3.3)* show the number of days (or months, etc. or percentage of time) in a year that any given flow has been exceeded during the period of record. The area under the curve shows the total volume of the discharge in the year, which provides the average yearly flow. Flows which were exceeded, say, 90% of the time give a first indication of the minimum flows during the driest period, while flows which are exceeded only, say, 2% of the time indicate the flows during floods and the duration of floods. The information can be dealt with just numerically but the curves in graphical form give a valuable visual indication of the characteristics of a river. For instance, flat curves indicate steady, continuous discharge, often in large rivers. A very curved form tends to show 'flashy' rivers (Fig. 3.3(a)). Zero discharges, of course, show that the river is not perennial. The curves can be prepared for monthly, seasonal, yearly or any other periods, while the complete record plotted on one chart shows the 'long average' behaviour of the river. The area of the curve under any particular flow shows the volume of the flows which could be utilized by diversion through a conduit of that maximum capacity. Thus from the curve can be prepared a discharge utilization curve (Fig. 3.3(b)) to show the percentage of the annual flow which could be diverted for a scheme of any given maximum capacity on a run-of-river basis, i.e. without reservoir storage. The curves can easily be converted to power duration and energy utilization curves by multiplying the discharges by a function of the potential head, including allowances for head losses and efficiency of the hydropower station.

- *Mass curves (Fig. 3.4)* are compiled by plotting the flows in the river for the whole period of record at a possible dam site on a cumulative basis as ordinate, against time. As in the case of the flow duration curve, this can be done for daily, 10-day or monthly flows, or flows of any period. For prefeasibility studies, a monthly time module should be sufficient. To prevent the cumulative curve from ascending ever higher and higher off the page, it is convenient to plot the flows as the excess above or deficiency below the estimated average value. The gradient of the curve then represents rates of flow above or below the estimated average. Cumulative discharge flows can then be plotted in the same way, as differences from the estimated average inflow, a constant gradient representing a constant rate of discharge. If the inflow and outflow curves are superimposed, starting together at a crest of the inflow curve, the maximum height between the two curves represents the volume of storage required to maintain the discharge plotted. Hence the maximum height above the inflow curve of a straight line which is tangent to any two crests of the inflow curve, represents the storage required over that time interval to maintain the discharge rate represented by the straight line. Spillway discharge during floods is represented by periods when the inflow curve is above the discharge curve. It can be seen that the greatest discharge rate overall is represented by a straight line, otherwise what goes out of a reservoir is being shown as greater than what comes in. The year-over-year storage required to maintain the greatest possible average outflow (equal

Selection of potential dam sites and conceptual schemes | 33

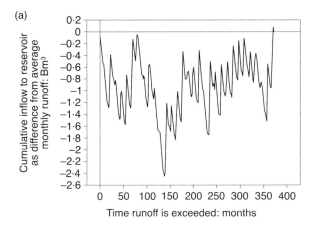

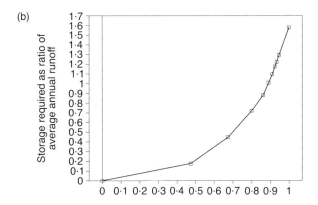

Fig. 3.4. (a) Unitized cumulative monthly inflow. (b) Storage required to yield continuous discharge as factor of monthly inflow. K is the factor of the average annual runoff required as storage to yield a continuous discharge as a given factor of the monthly average inflow C e.g. in the plot K indicates that a storage of about 1·6 × average annual runoff is required (on average) to provide a continuous yield equal to the monthly average inflow

to the average inflow) over the period of record (assuming no spill to waste during floods) is represented by the greatest height above the inflow curve at any point of a horizontal discharge line which is tangent to the highest crest of the inflow curve (a line a little bit lower than the average inflow line shown on Fig. 3.4(b)). Note that 'outflow' must include losses due to waste, leakage and evaporation.

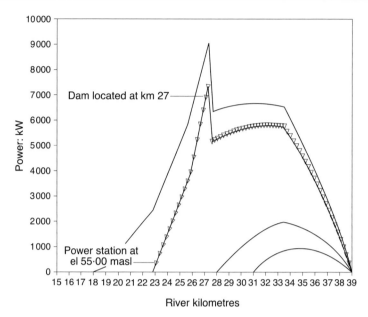

Fig. 3.5. Hydropotential curve

In this way it is quick and easy to obtain first estimates of the utilization of its discharge which can be obtained from a river, and the reservoir capacity and conduit capacities which will be required for a project to achieve any proposed output capacity.

If contoured mapping permits, the storage capacity against water level curves for promising reservoir sites can now be prepared by planimetering the area of each contour above the possible dam site and calculating the volumes above river bed level. The maximum storage capacity to be developed at a site is often decided by that height of dam, to the level above which the topography flattens off, which would require an extravagant increase in the length of dam, or above which valuable land and property would be inundated. These maximum storage capacities can be marked at the appropriate points on the longitudinal profile of the river.

If hydropower development is needed, potential hydropower outputs along the river profile can be estimated and presented by selecting a potential power station site as far downstream as might be contemplated (Fig. 3.5). From the maximum discharge capacity for each reservoir site under consideration and the head difference between the maximum reservoir level at each site and the river level at the common power station site, making some allowance for conveyance losses (say, 5%) and efficiency of generation (say, 85%), the potential average power output for each reservoir (i.e. intake site) can be estimated. By plotting the potential power output against the chainage of the dam site along the river, the intake site giving maximum power output is readily seen. By repeating the process for different possible power station sites further upstream, the combination of intake and power station site which would be likely to yield the greatest average power output economically can be anticipated, as shown in Fig. 3.5.

It should be noted that the appraisal of hydropower potential just described is for the purpose of preliminary choice of sites only. The optimum capacity of power plant to be installed at the station may be found to be much greater than the potential power output because the station may be required to operate for only a few hours per day to supply peak loads, in which case, using reservoir storage, the plant may run at up to, say, six or even more times the average output available over 24 hours. It might be running for only, say, 4 hours per day.

The preliminary layouts of the principal works and of alternatives proposed where appropriate, should then be prepared on the basis of experience and using data from comparable existing projects. Cost estimates can be prepared by comparison with the costs of similar works elsewhere, with adjustments to take account of any conditions in the two cases which are different. Adjustments should be made for varying freight and transport costs, exchange rate variations if in different countries, comparison of rates quoted in tenders for similar work in the two places, price escalation if there has been a time lapse between estimates, etc. In the case of hydropower projects, benefits may be calculated by estimating the cost of the least costly alternative form of generation which would otherwise have to be developed instead of hydropower. The present value of costs should be used in order to make a fair comparative evaluation of schemes with a varying proportion of capital costs and recurrent operating and maintenance costs. The present value of benefits will lead to the estimation of preliminary but indicative benefit/cost ratios.

Choice of dam type

The main types of dam are now described.

- *Embankment dams*, which may be:
  - earthfill, with inclined or vertical impermeable clay core, or homogeneous fill type (Fig. 3.6) or
  - rockfill with inclined or vertical impermeable clay core, or central or upstream asphalt membrane, or central or upstream concrete membrane.

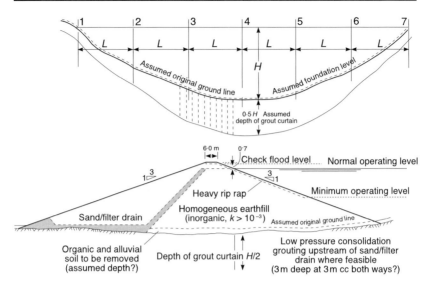

Fig. 3.6. *Schematic design of earthfill embankment dam suitable for pre-feasibility planning*

- *Concrete dams* which may be one of the following types.
  - Mass concrete dams which may be of solid gravity type, or hollow with massive buttresses, or curved arch gravity type, offering the mobilization of some degree of arch action. The concrete can be of conventional construction or it can be roller compacted concrete.
  - Arch dams of relatively slim cross section (like Vaiont Dam, Fig. 3.1), ideally with upstream and downstream faces separately curved both in plan and in elevation.
  - Hollow buttress dams of the so-called Ambursen type with reinforced concrete upstream and downstream slabs and reinforced concrete buttresses.
  - Multiple arch dams effectively supported on buttresses across a valley.
- *Other variations* based on the above types.

A reasonable principle in deciding the suitability of a type of dam for a site is that the material in the foundations in contact with the dam should not be of lower strength than the material in the dam itself. Thus a sound rock foundation may be able to support any type of dam but an earthen foundation will only be suitable for an earthfill dam (on the principle that, at the dam–foundation interface, the stresses in the two materials are equal and opposite). A narrow valley with sound rock river bed and abutments is likely to be suitable for a concrete arch dam, while the placement of large volumes of fill in embankments requires access for and the manoeuvrability of heavy earth moving plant, which calls for wider valleys with flatter slopes.

Concrete dams need to be founded on good foundations, preferably sound rock, so the need to remove a considerable depth of inferior material above sound rock would incline the choice towards a fill dam. Leakage past a dam needs to be reduced to a very small amount, so the dam needs to be considered in relation to the grout curtain and/or the cut-off trench or core wall which would be required to prevent leakage paths under or through the dam or on its flanks, and the depth to which the cut-off would need to be taken. Heavy leakage could not only represent a loss of valuable water but also the potential for an unacceptable increase in leakage by causing internal erosion.

Buttress dams have some advantages over solid concrete gravity dams in that the buttresses spread the load in an upstream–downstream direction, resulting in lower intensities of loading and excellent resistance to overturning. The spaces between buttresses offer shorter paths for drainage and relief of uplift pressures. Figure 3.7 shows two examples of diamond headed buttress dams, Itaipu (Brazil) and Wimbleball (UK). A principal alternative to the diamond head is the round head type. Shuttering (formwork) for buttress dams is clearly more complicated, extensive and costly compared with that for solid gravity dams. Lateral resistance to earthquake vibrations is also less inherent.

The type of dam required needs to be decided with knowledge of those materials which would be economically available locally, without high transport costs: the supply of cement and aggregate for concrete, or sites for quarrying sufficient sound rock for rockfill; sites for borrow pits for earthfill in sufficient quantities; timber or steel for formwork for concrete and the skilled carpenters to erect it — considerable quality control and accurate fixing are required, especially for arch dams with complex curvature. *Can all these be obtained economically?*

Roller compacted concrete (RCC) is a material which has been developed after research over many years and has more recently increasingly displaced conventional concrete for dam construction. In making RCC, a large proportion of the cement required for concrete is replaced by pozzolanic materials, often fly-ash. Some four design philosophies have been developed: the lean RCC dam, the rolled concrete dam (RCD) and the medium and high paste content RCC dams. The cementitious contents (i.e. Portland cement and pozzolan) have varied from less than $99 \, kg/m^3$ in the first case (RCC) to over $200 \, kg/m^3$ in the last (high paste content) and the pozzolan contents from less than $40 \, kg/m^3$ in the first case to over $120 \, kg/m^3$ in the last.

38 | *Reservoir engineering*

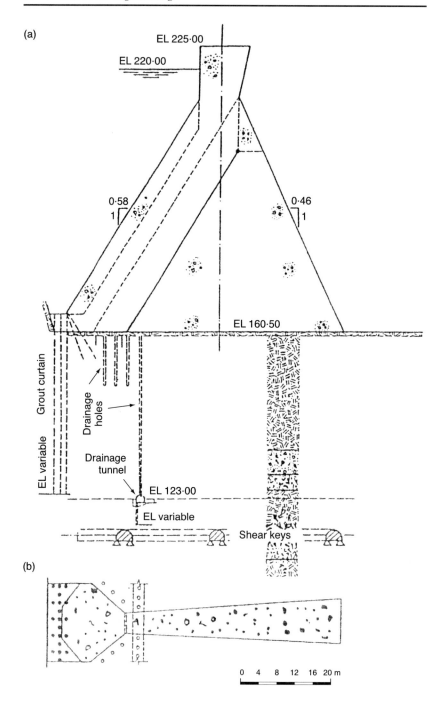

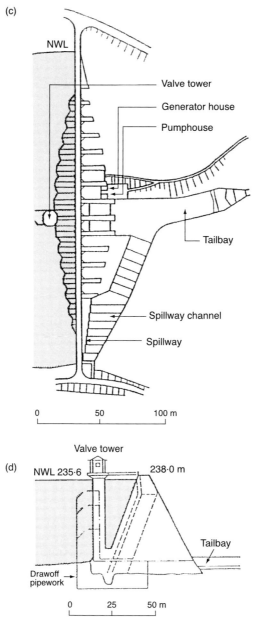

Fig. 3.7. Diamond headed buttress dams. (a) and (b) show the section and plan respectively of the Itaipu Dam, Brazil (maximum height, 196 m). (c) General arrangement and (d) section through dam and valve tower of Wimbleball Dam in the UK (height 63 m; upstream slope of buttress 1:0·383)

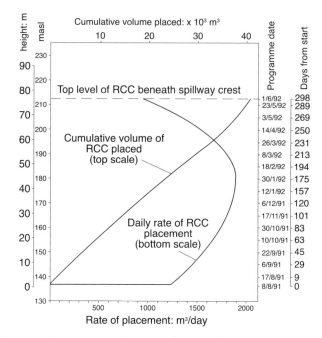

Fig. 3.8. Rate of production of concrete for a particular RCC dam

Bulk handling and continuous mixing and placing methods are used for RCC, more akin to those conventionally used for placing fill than placing concrete. Joints and formwork are minimized and rates of progress can be far greater than for conventional concrete dams. Rates of placement of RCC in dams have reached average rates of 2000 m^3 per calendar day over periods of up to 10 weeks and maximum rates of up to 10 000 m^3 per day (Fig. 3.8). Generally RCC dams can be constructed at a rate of approximately 2 to 2·5 m of vertical height per week but sometimes as much as 5 m or more per week (Fig. 3.9). For comparison, remarkably, about 3300 m^3 per day on average of conventional concrete was placed in a double curvature concrete dam (perhaps the most time consuming type of dam) in South Africa in 1994. Consequently in many situations suitable for concrete dams, RCC is the most economic and rapid form of construction. The pozzolanic materials used can be of a wide variety (e.g. from natural sources or from spent fuel ash from thermal power stations) but they must be suitable, of consistent quality and well controlled in production.

My view is that a rockfill embankment construction with upstream membranes (Fig. 3.10) has three important advantages over internal cores.

Selection of potential dam sites and conceptual schemes | 41

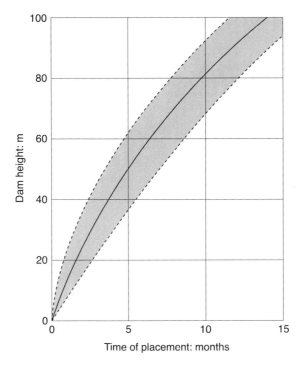

Fig. 3.9. Speed of construction of RCC dams

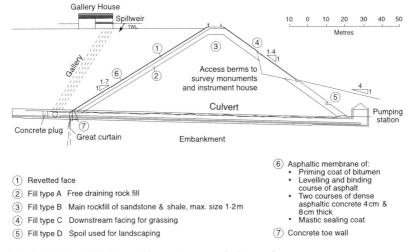

① Revetted face
② Fill type A Free draining rock fill
③ Fill type B Main rockfill of sandstone & shale, max. size 1·2m
④ Fill type C Downstream facing for grassing
⑤ Fill type D Spoil used for landscaping

⑥ Asphaltic membrane of:
  • Priming coat of bitumen
  • Levelling and binding course of asphalt
  • Two courses of dense asphaltic concrete 4 cm & 8 cm thick
  • Mastic sealing coat

⑦ Concrete toe wall

Fig. 3.10. Rockfill dam with upstream asphalt membrane

(a) The construction of the membrane can be carried out in one or two stages, when the fill is completed to sufficient height, quite independently of placing the fill. The placing of central cores and fill may be affected by weather conditions and either may delay the other. Upstream membranes take a relatively short time to place, which can usually be programmed for a dry season *after* the fill has been placed.
(b) The water load is applied by the membrane to the whole volume of the dam, not only to the downstream shoulder. The upstream shoulder is not saturated. This geometry is technically attractive and usually results in a smaller volume of fill in the dam and a shorter construction period.
(c) An upstream membrane is much more easily accessible for inspection and repair (should it be necessary) than is a buried core.

However, some engineers favour buried cores.

With embankment dams, it is customary to have a separate spillway for flood discharge either on one flank or over a separate saddle in the rim of the reservoir. This is because of the high cost of supporting the spillway (if within the dam) and its side walls, and of ensuring that the fill material is protected from erosion. Preventing seepage and erosion at the interface of a concrete spillway with fill material in the embankment is also a difficulty. With concrete dams, including RCC dams, floods can be discharged over an open spillway over the crest of the dam or through a gated spillway in the dam itself. This disadvantage of fill dams helps to make concrete and RCC dams competitive.

It may be found that, if different types of dam are feasible at a site, there may not be decisive cost or technical advantages of one well designed type over another, particularly at the prefeasibility stage, because of the preliminary nature of the designs and cost estimates: the margin of error may be greater than the estimated cost difference between two types. It is important, however, to try to home in on the best type in order to restrict site investigations to those needed for that type and to avoid abortive time and costs in making alternative designs. Therefore all significant factors should be taken into account in making recommendations, which should be discussed with the employer before they are finalized.

Choice of size of dam

It is worth noting here that a wealth of practical information is contained in the publication *Design of Small Dams* [2]. Generally this volume is intended to cover structures with heights above the stream bed not exceeding 50 ft (not exceeding 20 ft for concrete dams on pervious foundations) but much of the contents are relevant to larger dams, and cover associated structures and the engineering and principles involved.

During the prefeasibility stage it is necessary to prepare preliminary layout drawings and cost estimates without resorting to specific design work. The estimates can be prepared by experience and comparison with analogous schemes. Usually approximate estimates of quantities of the principal elements of construction, including excavation, must be made. Unit cost rates should be applied on the lines just described for hydropower projects. It is important to ensure that sufficient allowance is included for contractors' overheads and profit which can add a mark-up of between 25 and 100% to the basic cost. An allowance must be included for temporary works, such as river diversion. A mark-up needs to be added for general items which could include the provision of a performance bond, contractors' third party and general insurance, temporary construction camp and access for construction. For these, a mark-up of the order of 10% would be quite usual. At this stage, benefits may be estimated on the basis of the value of the outputs to the owner, or, alternatively, as the financial benefits of the sale value of the outputs, represented by the present value of avoided costs (i.e. the costs which would be incurred if the outputs were obtained from some other alternative development, including the costs of building the alternative). To obtain present values, a discounted rate of return has to be agreed. In the author's view, this should be some 4% less for dam and reservoir projects than for other alternatives which may be subject to escalating recurrent costs and which have shorter lives. The subject of costs and benefits is dealt with in Chapter 14.

The height of dam can now be optimized in a preliminary way by estimating the cost of dams to at least four different crest levels, and substituting each in turn in the benefit/cost estimates. An optimum should be found because a low height of dam provides little storage and hence low firm output, while an excessive height incurs very high cost. If there is not sufficient current demand to warrant building the dam to the maximum storage capacity which could be justified for the site, the alternatives of over-building the dam in one stage or postponing the raising of the dam to the full height to a later stage should be costed and put to the client. A decision may depend on the economic viability of the project and on the finance available early on. Division of the work into two stages reduces the investment at first but could incur the duplication of cost of mobilization of contractors, and additional costs in design and construction later, as well as additional time spent on construction overall and perhaps some technical difficulties in making the second stage of the dam integral with the first.

References

[1] Charles, J. A. *et al.* (1991). *An Engineering Guide to Seismic Risk to Dams in the United Kingdom.* Building Research Establishment, Garston, UK.
[2] US Department of the Interior, Bureau of Reclamation (1974). *Design of Small Dams.* Denver: US Government Printing Office, Denver, Colorado, 3rd Edition (1987).

4. Investigation of selected sites and geological studies

Site investigations
In Chapter 2, reference was made to the assembly of geological data and in Chapter 3 to the appraisal of that data. There was reference to site reconnaissance and to prefeasibility studies, which would produce recommendations and (perhaps) terms of reference for investigations which would be required for detailed feasibility studies. The point has been made that one of the most important early considerations is the review of data and ground reconnaissance by an experienced engineering geologist. The investigations, however, will be incomplete unless and until there are sufficient sub-surface investigations. However much investigation is carried out, it cannot reveal a complete and detailed portrayal of all the geological conditions which may be important and that is why the understanding and perception of an experienced engineering geologist is so important in the interpretation of surface conditions and of data on sub-surface conditions. Boreholes cannot reveal continuous information in three dimensions even at excessively expensive close centres, except by interpretation. A fault that outcrops on the surface cannot easily be located at a depth of tens of metres because of deviations of the fault from a plane, or of deviations by only one or two per cent from a straight line of the drill hole (1% of 10 m is 100 mm). Nevertheless, the geologist cannot dispense with sub-surface information and is best placed to decide the nature and strategic siting of such investigations. One or more boreholes may show a joint or fracture in the core, perhaps with clay partings or gouge. But how continuous in the ground is that small, isolated feature in a core? The Author has come across a site where such slight evidence represented extensive planes of weakness, along which sliding of the foundations could occur. The foundations had to be post-tensioned with cable anchorages to improve their resistance to sliding. The geologist foresaw this but the geologist's concerns were dismissed until the excavations for

the dam revealed the extensive and continuous planes of weakness to shear. At another site, geological reconnaissance revealed evidence of serious slope instability and resulted in a rapid geomorphological survey, which resulted in the site intended for the dam being moved some 600 metres downstream, although by that time investigations and design on the original site were quite well advanced. The following lists some tools that are available for site investigation.

(a) *Geological survey* — comprising field mapping to amplify existing surveys and to plot and interpret surface features (with the aid of existing sub-surface data), and to locate potential sources of sufficient quantities of materials for construction.
(b) *Geomorphological survey* — to determine the past processes of land formation (geological/tectonic/alluvial/colluvial/sedimentary/etc.) and their effects on present stability, as well as to predict future stability, rates of erosion, and sedimentation in proposed reservoirs.
(c) *Trial pits and trial trenches* — to expose and to examine visually the near-surface strata and bedrock, if sufficiently shallow to be reached by trial pits.
(d) *Boreholes* — to carry out tests and to determine the depth to bedrock in the case of rotary wash drilling or percussion drilling, or to recover samples of overburden and rock in the case of core drilling.
(e) *Trial adits* — to expose the strata to visual examination and possibly to allow testing on a large scale.
(f) *Geophysical surveys* — usually using seismic or electrical resistivity techniques, to assess the variation in level and nature of strata between boreholes in continuous lines at much reduced cost compared with numerous boreholes. More recent developments in ground radar are extending the utility of this technique. However, none of these techniques is reliable in all circumstances and their interpretation cannot be relied upon without correlation with other, direct, exploration methods.
(g) *Obtaining samples* from trial pits and boreholes, etc. for laboratory testing.
(h) Carrying out *in situ tests* in trial pits, boreholes and adits.
(i) *Installing instruments in trial pits, boreholes and/or adits* — notably to measure deformations and in situ stresses.
(j) *Installation of piezometers in boreholes* — to monitor water levels and pore pressures.

Clearly the above is not an exhaustive list but it summarizes methods in frequent use. Attention will now be drawn to some points worth noting for each of these tools.

Methods
Geological survey
The most important consideration has already been mentioned: geological surveys, if required, should be undertaken by an experienced and perceptive geologist who will need the assistance, preferably at the time, of a small survey team to locate on maps the features which are to be recorded, and which should be marked on the ground. If the features cannot be located on adequately contoured maps or if they occur on steep slopes, their levels will also need to be surveyed. Two of the geologist's main objectives will be, firstly, to locate possible planes of weakness under the dam and/or in its abutments and/or in the slopes above the proposed reservoir and, secondly, to locate possible paths of leakage under the dam, through its abutments and from the reservoir, either in a downstream direction or into neighbouring valleys. The geologist will also be observing features subject to erosion, and should be able to locate and assess suitable sites for rock quarries and areas for borrowing materials for construction. To these ends, the geologist will need to assess the stratigraphy, rock types, structure, weathering, faulting, jointing and soils.

Geomorphological survey
If required, geomorphological surveys should be carried out by a specialist or specialists in collaboration with the geologist and will commence with reconnaissance and study of topographic mapping, study of aerial photographs by stereography and study of satellite photography. The studies should enable the catchment to be divided into zones of similar land forms, steepness, soil cover, vegetation and land use, as in Fig. 4.1. From these parameters and those of rainfall intensities and wind speeds and direction, and topographical surface gradients applicable to each zone, estimates of the contribution of each zone to suspended sediment load and bed load in each tributary, and hence in the main river and into the reservoir, can be made. A case history [1] is described here. Where possible, the estimates of sediment load should be checked and if necessary adjusted, by comparison with the results of in situ sampling and testing and with rates of sedimentation which may have been observed locally. The index on Fig. 4.2 shows volume of erosion per km^2 per year for catchments worldwide and the variation is an order of magnitude of ten times for a given size of catchment — not very surprising, considering the likely variation in the nature of the rocks and soils. Much more data on rates of sedimentation have been reported [2].

The fundamental causes and mechanism of slope instability should be determined if possible. Features of potential instability should be located on maps and the magnitude of potential movements within

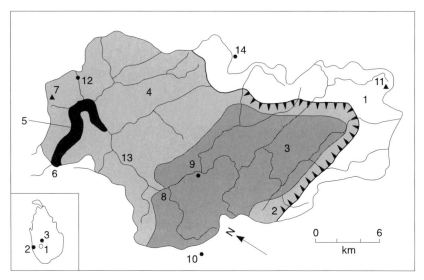

Fig. 4.1. Location map and sub-division of the Kotmale drainage basin: (1) high surface, (2) escarpment, (3) low surface, (4) gorge region, (5) reservoir site, (6) dam axis, (7) Peacock Hill 1583 m, (8) Saint-Clare Falls, (9) Talawakele, (10) Hatton, (11) Totapolakanda 2380 m, (12) Helbodde, (13) Kotmale river, (14) Nuwara Eliya
Inset: (1) Kotmale project, (2) Colombo, (3) Kandy

the life of the reservoir should be estimated, with emphasis on those movements which could descend into the proposed reservoir, together with estimates of their terminal velocities, as shown in Fig. 4.3: this is in the nature of a map of horrors and shows the limit of the reservoir, and the limit of rock fall runouts.

Trial pits and trial trenches

Trial pits are usually excavated to the minimum dimensions necessary for man-access, with sufficient clearance for revetment to support the excavated sides. Trenches may be excavated manually, or mechanically by dozer or backacter. To limit expense, depths usually do not exceed about 3 m but can be taken to 5 m or even deeper. The excavations permit large-scale visual examination and should be logged by an experienced geologist. Bulk disturbed samples should be taken and be bagged or boxed and well labelled. Natural moisture contents, particle size distribution and Atterberg Limit tests are normally carried out on representative bulk samples. In situ density tests in granular soils and Standard Penetration Tests (SPTs) and undisturbed samples (100 mm diameter) in cohesive soils, should be taken for measurement of in situ dry densities and (in cohesive materials) for classification tests and unconfined compression tests. For cohesive fill destined for dam construction, triaxial testing in the laboratory is likely to be required. Dry densities and permeabilities can also be measured in situ on larger and more representative samples that cannot be easily transported to the laboratory. In the areas of intended borrow pits, compaction tests, classification tests and determination of California Bearing Ratios (CBRs) should also be carried out to establish standards of compaction achievable in the dam and in road embankments.

Triaxial testing will normally be by the 'consolidated undrained' method with measurement of pore pressures to investigate conditions during construction and to investigate stability of the upstream face during rapid draw down of the reservoir, which are usually conditions with durations too short to allow full drainage. The long-term stability of the dam under drained conditions is usually less critical. It may be noted that where it is necessary to test rockfill for dams, it is possible exceptionally (and expensively) to carry out triaxial and oedometer (settlement and deformation) tests on large specimens of 150 mm, 300 mm and even up to 1·0 m in diameter in order to include more representative particle sizes. Such samples would be obtained from quarries or from trial drilling and blasting.

Boreholes

If the overburden is to be removed, rotary wash drilling or rotary percussion drilling is relatively rapid and inexpensive but to be avoided if continuous information is needed on the nature of the strata because of the disturbance caused to the ground and because undisturbed samples can be obtained only by stopping drilling and recovering such samples by fully approved methods. While drilling, fines content in the soils is likely to be lost in the wash water. In soft ground, including in granular soils, holes will usually need a steel casing. This entails drilling at a larger diameter to allow insertion of the casing as the hole progresses. Due to frictional resistance from the ground, it may be necessary to reduce the diameter of the casing and of the drill, at some depth. If this is foreseen, the hole should be started at a larger diameter than required finally.

For diamond core drilling in overburden (which is expensive due to the need to case the holes for support, in order to avoid collapse of the perimeter) or for diamond drilling in rock, ideally holes should not finish at less than about 100 mm diameter (denoted by HWF or HWG) to obtain cores of minimum 75 mm diameter (BS 5930: 1981 [3]). The best practice is to use triple core barrels, which prevent disturbance of the core by contact with the rotating bit, and cause

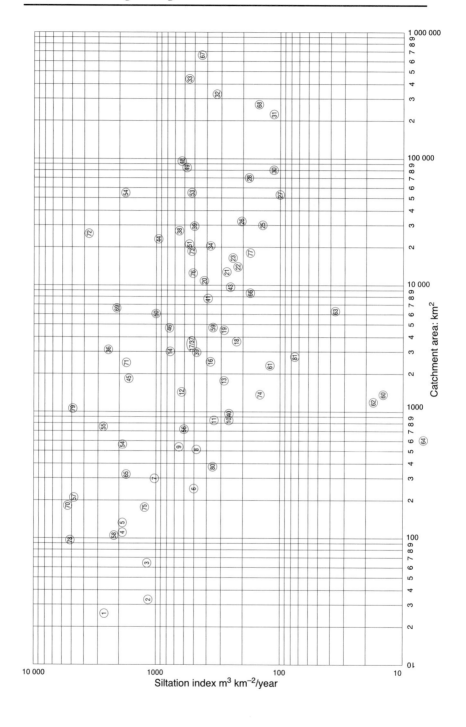

Investigation of selected sites and geological studies | 51

Ref. No.	Name of reservoir	Ref. No.	Name of reservoir	Ref. No.	Name of reservoir
1.	Lake Duncan, Oklahoma, USA	28.	Elephant Butte, N. Mexico	55.	*Shihmen, Taiwan
2.	Guthrie, Oklahoma, USA	29.	Denison Dam, Texas, USA	56.	*Ku Kuan, Taiwan
3.	Lake Clinton, Oklahoma, USA	30.	Bou Hannifia	57.	*Wu Sheh, Taiwan
4.	Mankla	31.	Orib	58.	Taipu, Taiwan
5.	Namiz	32.	Mead, Nevada, USA	59.	Roznow, Poland
6.	Chambon	33.	Boulder Dam, Colorado, USA	60.	*Turawa, Poland
7.	Morena	34.	*Matatila, Uttar Pradesh, India	61.	*Otmuchow, Poland
8.	San Gabriel No.2, California	35.	Wadi Jijan, Saudi Arabia	62.	*Pilchowice, Poland
9.	Gibraltar, California	36.	*Ramganga, Uttar Pradesh, India	63.	*Pernegg, Austria
10.	Sautet	37.	Pitesti (V Arqes), Romania	64.	Steyerdurchbruch, Austria
11.	Fodda	38.	Tungbhadra, Kanataka, India	65.	*Wetzman, Austria
12.	Alt Aaddel	39.	Karangkates Dam, Indonesia	66.	*Roxburgh, New Zealand
13.	Lalla Takerkoust	40.	Capivari Dam, Brazil	67.	*Sanmenxia, China
14.	Bridge Port, Texas	41.	Passo Real, Brazil	68.	*Quingtongxia, China
15.	Iblanica	42.	*Roseires Dam, Blue Nile,	69.	*Demirkopru, Turkey
16.	Lake Dallas, Texas, USA	43.	*Kamburu Dam, Kenya	70.	*Buldan, Turkey
17.	Serre Poncon, France	44.	*Gandishagar Dam, India, M.P.	71.	*Kemer, Turkey
18.	Aqua Fria, Arizona, USA	45.	*Mayurakshi Dam, India, W.B.	72.	*Hirfanli, Turkey
19.	El Kansera, North Africa	46.	*Girna Dam, Makarashta, India	73.	*Seyhan, Turkey
20.	Alamogordo, N. Mexico	47.	Bhavani Dam, India	74.	*Forggensee, Bavaria, Germany
21.	Roosevelt Dam, Arizona, USA	48.	*Sriramsagar, India, A.P.	75.	*Ringlet, Malaysia (1996)
22.	Zaouia M'Ourbaz	49.	*Hirakud Dam, Orissa, India	76.	*Mellegue, Tunisia
23.	Conchas Dam, N. Mexico	50.	Maithon Dam, Bihar, India	77.	*Sidi Salem, Tunisia
24.	Imfout, North Africa	51.	*Nizamsagar Dam, India, A.P.*	78.	*Lebna, Tunisia
25.	McMillan, N. Mexico	52.	*Shivasisagar Dam, India	79.	*Marguellil, Tunisia
26.	Gila	53.	Bhakra Dam, India	80.	*Cadore Valley, Italy
27.	Mechra Klila, North Africa	54.	TenKee, Taiwan	81.	*Matahina, New Zealand

Note: The source of the above data was in places indistinct. Therefore, without further verification, names and figures should be taken as only a guide. The author and publishers would welcome corrections and additional data from readers for use in future editions. As noted in Chapter 4, extensive data has been presented in *Evacuation of Sediments from Reservoirs*, White. R, Thomas Telford Ltd, London, 2001. Where data from that source has been used above, it is denoted by *.

Fig. 4.2 Siltation Index

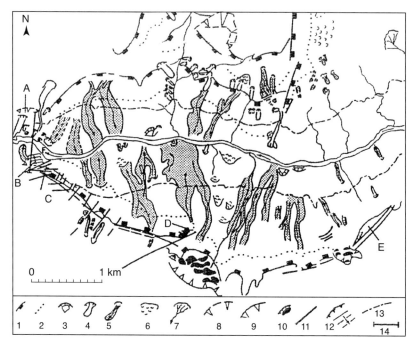

Fig. 4.3. Extract from the geomorphology/landslide hazard maps for the Kotmale valley: (1) cliff face susceptible to rock fall, (2) predicted maximum distance of rock fall runout, (3) planar translational slide, (4) debris slide, (5) mudslide, (6) active soil creep, (7) eroding headwater area, (8) ancient landslide scar, (9) active or recent landslide scar (deep seated), (10) slumps, (11) tension cracks and critical major discontinuities, (12) major deep seated block slide, (13) reservoir limit, (14) dam axis, (A) unproven landslide? (B) major landslide area, (C) buttress showing incipient failure, (D) rock spur, incipient instability, (E) potential major deep seated landslide bounded by tension cracks, (F) Kotmale river

the core to enter a tube in which it is enclosed on withdrawal to prevent loss of cored material. In this way, continuous cores should be obtained which disclose joints and fractures in the rock. Cores should be well labelled, stored in full-length core boxes, logged by a geologist and stored in weather-proof conditions. The location of the borehole and depth of cores should be marked on the box. A useful characteristic of the soundness of the rock, the Rock Quality Designation (RQD) should be determined and logged for each length of core by counting the number of pieces of intact core of 100 mm or more in length, as a percentage of the length of core. The value is usually established for each 2 m run of core. The core should be of not less than 50 mm in diameter, drilled with double barrel diamond drilling equipment. The originator of the parameter, Professor Don Deere, considered that RQD values of less than 25% represented very poor rock, 25–30% poor rock, 50–75% fair conditions, 75–90% good and 90–100% very good. The reliability of the RQD is very much a function of the quality of the drilling. Samples of rock core can be subjected to unconfined compression testing and testing to determine the void index, which is a characteristic of rock quality and strength: the void index is the weight of water absorbed by a specimen of dried rock material of approximately 200 g in weight, immersed in water for 12 hours, as a percentage of the weight of the sample after drying for 12 hours at 105 °C. There is a minimum void index for any particular rock, usually less than 1·0. The nearer the value is to 12, the nearer the material is to the rock–soil boundary. For rock materials with void index greater than 12, immersion in water is likely to produce disintegration. If required by the geologist, thin slices of the cores can be cut for petrographic examination. The geological age of rock can be determined by carbon dating or other methods, to determine, by age, the sequence of deposition.

Trial adits

The excavation of trial adits allows continuous visual examination and may avoid the need for numerous deep boreholes. Behaviour of the ground during and after excavation can be observed. Often this is most valuable on the steeply sloping abutments of dams or to explore specific features in the foundations or the possible sites for tunnel portals, where the occurrence of weathered rock makes tunnelling conditions uncertain. The observation of safety conditions and adequate ventilation for personnel visiting and working in adits is very important. Reference is made shortly to tests which may be carried out in adits.

Geophysical surveys

Of the geophysical methods, seismic surveying is generally preferred as being more reliable than electrical resistivity. Both involve measurements on the surface set up at intervals along lines (preferably between boreholes, since correlation with borehole results is essential). Validation by boreholes drilled subsequently, or intersected but not used as data for the survey, also adds to confidence in the results. Seismic surveys depend on the travel time of dynamic waves reflected by underlying strata and with different angles of incidence and different velocities in different strata. The seismic impulse is best created by small explosive charges but can be generated by the impact of a hammer blow. Cross-hole seismic surveys may also prove valuable. Electrical resistivity surveys interpret the depth of varying strata by measuring the electrical resistance through the ground between transmitting stations on the surface, at a range of distances apart, and subjected to a range of difference in electrical potential. Both seismic and resistivity surveys are best carried out by specialists experienced in these. The effects of groundwater and irregularities in the strata demand great care in siting the observation points and in the interpretation of results. Ground water particularly may rule out the use of these techniques, and the feasibility of their use at any particular site should be assessed on the basis of the known geology, by an experienced practitioner before embarking on a survey. Allowance needs to be made for potential errors in the analysis of observations.

Obtaining samples from trial pits and boreholes for laboratory testing

Salient matters under this heading have been dealt with earlier, in relation to the sources from which the samples can be obtained. Only two further matters are noted here: first, that transporting, storing and testing samples can be expensive and time consuming and care must be taken to disturb them as little as possible during the journey, so whenever possible these actions should be restricted to representative samples, and duplicate samples of the same material from the same strata should, after logging, be kept only as long as needed for inspection and record on site. They should normally, however, be retained through the design stage, and for inspection and reference during tendering and construction, especially in case unforeseen ground conditions are discovered during excavation, leading to claims for extra payment by contractors.

Second, selected undisturbed samples of soil for testing and selected samples of rock cores liable to be altered by exposure to the atmosphere or likely to be selected for index tests, should be sealed by coating with wax, wrapping in polythene sheet and placing in sealed containers, as soon as possible after extraction from the ground.

What has been said here has been derived from the Author's experience but the National Accreditation of Measurement and Sampling (NAMAS) scheme 1989 has become the UK Accreditation Service (UKAS) and reference to ISO 9000 standard is necessary for its approval.

Carrying out in situ tests in trial pits, boreholes and adits

The modulus of deformation and compressibility of soft ground, fill or rock can be determined by large-scale plate bearing tests, the size of plate depending on the loading which it is required to test and the load which it is possible to mobilize by heavy weights and/or ground anchorages or reaction to a powerful hydraulic jack. The loading can be applied to the floor and sides in trial pits and adits, and to the soffit of adits. Similarly, the in situ shear strength of soils and rock (for example, along a suspected plane of weakness) can be determined by applying the loading from a jack to the face of a rectangular block of natural ground, which has been excavated carefully all around and on top, in a pit, above the plane of weakness, in a manner so as not to disturb the structure and support of the block. In such tests, reference datum points must be established which will not be disturbed by the tests, from which deformations can be accurately measured.

In situ permeability can be measured in pits or boreholes by falling or constant head methods, measuring the rate of fall in level of water in the pit or borehole, or the rate at which water has to be replenished to maintain a constant level. Permeability tests in piezometers are usually more reliable than down-the-hole or infiltration tests and are to be preferred when feasible.

In boreholes for reservoirs, it is good practice to carry out water pressure tests in rock in each different strata or at about 2 m intervals, whichever is more frequent. These tests determine the 'Lugeon' value of the rock, i.e. the water lost in litres per metre length of borehole per minute when subjected to a pressure of $10\,\text{kg/cm}^2$. The length of borehole in which the pressure is to be maintained is isolated from the rest of the borehole by expanding packers, installed by drill rods through which water is injected at the required pressure, the rate at which water is injected being measured. The Lugeon values can be derived by proportioning the results linearly to the standard units: to the length of hole subjected to test, to the pressure applied and to the duration of the test (the observed loss being reduced to give the Lugeon value if lengths, pressures and times of test are greater than standard). It is notable that the diameter of the borehole is not usually taken into account in assessing Lugeon values. More information can be obtained about the jointing and openness of joints in the rock by applying the pressure progressively and observing the rate of change in water losses. A C Houlsby has proposed such a method of refined interpretation [4] but this involves five 10-minute sub-tests at low, medium and peak pressures to assess permeability at the one location in the borehole, requiring interpretation of all five results — very time consuming and troublesome. A valuable paper on engineering of grout curtains to standards was also written by Houlsby [5].

Similarly, hydraulic fracturing tests can be carried out in boreholes between packers to determine the in situ state of stress in the rock. Water pressure in the borehole is increased until a sudden increase in the rate of loss of water indicates that the rock has fractured (or that

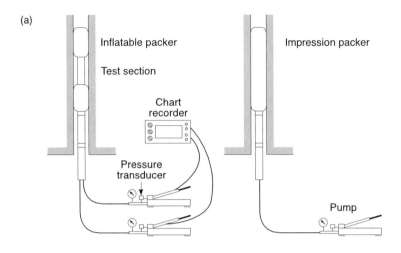

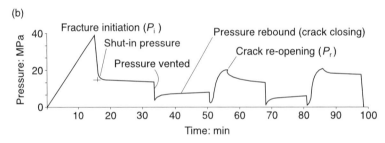

Fig. 4.4. (a) Diagram of hydraulic fracturing equipment. (b) Diagram of pressure record from hydraulic fracture test

joints have suddenly opened). The direction of crack can perhaps be predicted from knowledge of the jointing systems in the rock, or can be obtained by expanding a bulb coated with a plastic film in the borehole, to obtain an imprint of the crack. The orientation of the bulb has to be determined before its withdrawal. Portable 'micro' kits are available for hydraulic fracture testing (Fig. 4.4).

In situ grouting tests can be carried out in boreholes, to determine the rate at which cement grout can be injected with varying ratios of water to cement (from, say, 10:1 to 1:1) at varying pressures of up to, say, twice the hydrostatic head equivalent to the height of dam. These tests can give an indication of the quantities of cement which would be required to form an impermeable grout curtain below the dam or whether some grout, better penetrating than cement grout, would be required to reduce Lugeon values, to the order of 1 under concrete dams, 3 under earthfill dams and not more than 12 under rockfill dams (depending on detailed design and on risks of internal erosion in the foundations).

Pressure meter tests can be carried out in boreholes to measure the modulus of deformation of the ground when a bulb is expanded under pressure in the hole. This test is less suitable for rock of high modulus because of the relatively significant effect of compression of the bulb. Similarly a number of devices are available for applying pressure to the inside perimeter of a borehole and measuring the consequent deformation of the perimeter of the borehole.

In the ultimate test, a complete adit can be subjected to hydrostatic pressure tests, by instrumenting the adit with remote-reading strain and deformation gauges, constructing a watertight bulkhead at one end, with an access opening, and filling the adit with water under pressure. The objective would be to observe the performance of the rock in a tunnel under hydraulic pressure, in order to provide parameters for the design of tunnel and its lining. Such large-scale and costly testing would normally be justified in unusual rock conditions or for a very long and costly pressure tunnel, where optimization of the design would show very considerable benefit.

The Author would like to mention that full-scale trial compaction tests can usefully be carried out on site, to determine the acceptable amount of compaction needed, particularly of rockfill dams, using the compaction plant intended for use for construction. The tests should determine the optimum depth of layers, the number of passes of compacting plant needed and the densities achievable.

Installing instruments in trial pits, boreholes and/or adits

As shown diagramatically in Fig. 4.5, three-dimensional in situ stresses in rock can be estimated by installing by adhesive, three rosettes, each of at least three vibrating wire strain gauges, aligned at 45° to each other, on the circumference of a reduced diameter extension of a borehole, the rosettes being located at 120° to each other. The gauges are then overcored to release the in situ stresses, and the strain readings before and after overcoring are measured, the difference representing the released in situ strain. The stresses are derived from strains by measuring the modulus of deformation of the rock core in a laboratory. This method suffers from the necessary small scale of the strain gauges in relation to the scale of the rock particles and crystals and is liable to require the distribution of quite large errors, apparent if the number of gauges is sufficient to check whether the sums of pairs of orthogonal strains are equal, as required for uniform elasticity, and to check if strains in parallel gauges are recorded as unequal. Thus hydraulic fracture tests are preferred to

58 | *Reservoir engineering*

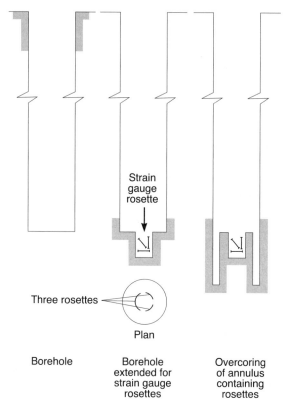

Fig. 4.5. Diagram showing the principle of stress measurements in a single borehole

tests using strain gauges, since they provide a more direct measurement and at larger scale.

In place of strain gauge rosettes, a single, large-diameter photoelastic stress cell can be installed in a borehole. This, too, can be overcored but estimating the stress distribution in the plane transverse to the hole requires viewing from the surface. The method is not always reliable. Stresses can also be measured by installing a rosette of four vibrating wire gauges on the flattened end of a borehole. This method is sometimes called 'the door-stopper method'. Both these methods suffer from requiring installation of gauges in at least two boreholes ending near the same point, if it is considered necessary to estimate the triaxial stress distribution.

Deformations in trial pits and adits can be measured by installing extensometers to measure deformations over a considerable base length, or across the width and/or height of the excavation, between fixed datum points, usually by means of an Invar wire fixed to one datum point and drawn by a plumb bob over the other.

Opening or closing of cracks and joints can be measured by installing some form of crackmeter: by stainless steel pins on either side of the joint (three pins at the points of a triangle at about 60°, in order to be able to calculate shear movements as well as movements transverse to the crack); by measuring the distance between pairs of pins with a micrometer gauge; or by vibrating wire gauges of sufficient base length and suitable design to cover the anticipated magnitude of movements. However the distances are measured, it is necessary to install the gauges in three directions to get a complete picture of movements in a plane.

The above account is not exhaustive since numerous types of deformation and settlement measurement gauges such as inclinometers to measure changes in slope in boreholes due to ground movements, electro-levels, load cells and other types are available. Many of these have more of a part to play in monitoring the behaviour of a dam than in investigations for its design. An excellent text is '*Geotechnical Instrumentation for Measuring Field Performance*' by Dunnicliff [6].

The installation of piezometers in boreholes

This facility is required for monitoring changes in groundwater levels, for example if affecting excavation during construction, and/or to provide information on the pressures liable to cause water inflows into tunnels during construction, and/or to monitor the changes of saturation level, and changes in pore pressure in embankment dams. For good results, well-proven, successful designs should be employed and carefully installed (Fig. 4.6). The simplest method is the direct measurement of water level in a hole with a perforated casing, encased in permeable sand. The water level is best detected electrically, when contacted by an electric-current-transmitting plumb bob. Similarly, readings can be transmitted for remote reading. Water levels in a sizeable piezometer hole containing water are not, however, sensitive to fairly rapid changes in pore water pressure and possibly take a long time to respond, except in very permeable soils. If the perforated casing or sand surround penetrates more than one water table, the results will be unreliable. For a more rapid response, Casagrande piezometer tips can be used, sealed into a short length of the hole. These are connected to the surface via a standpipe and dipped in the same manner as described. Alternatively, Bishop-type piezometers may be installed in the same manner as the Casagrande type and connected to mercury manometers or Bourdon tubes by small-diameter

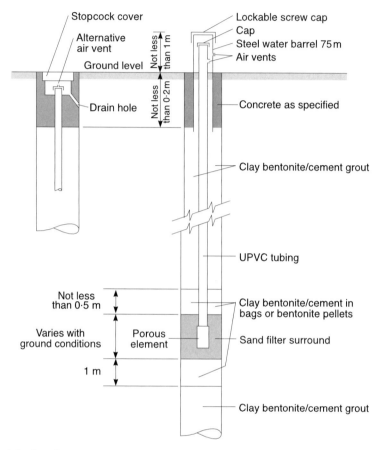

Fig. 4.6. Standpipe piezometer

hydraulic tubing filled with water, which must be de-aired. The measurements can be taken in a control centre, the centre being placed strategically for the measurements, in relation to the gauge positions. The manometers will not measure drops in hydrostatic head of more than about 7 m below the gauge level. The set-up and use of hydraulic lead piezometers is rather cumbersome and time consuming but the gauges are of good reputation, long-lasting and reliable if properly de-aired.

Other types of piezometers include electrical piezometers and pressure cells (a good type transmits the pressure on one side of a small membrane by change in volume of nitrogen gas on the other side, the volume change being measured by a transmitting extensometer, moved by a piston). Alternatively pressure gauges may be installed directly on the piezometer tube, providing the gauges are below the piezometric head.

More than one piezometer can be installed in a borehole by sealing each piezometer cell in a length over which it is desired to measure pore pressures. For example, as shown in Fig. 4.6, in a short length of hole, the hole above and below each is filled with bentonite (clay)–cement mix, isolating each length containing a piezometer from the remaining lengths of the hole. Piezometer boreholes must be covered with removable caps to prevent anything falling in.

Conclusion

The investigations and methods described in this chapter can be used throughout studies and designs of reservoirs in a progressive manner, the minimum investigations for preliminary purposes being undertaken in the prefeasibility stage (to confirm the feasibility of the concept and preliminary cost estimates) to an accuracy perhaps of the order of ±20%. In the feasibility stage, accuracies of estimates of ±10% or better will be looked for, and during design they should be improved to ±5% or better. It follows that parameters required for design should be of reliabilities in keeping with these margins of error. If anything, the construction and operation of reservoirs should be monitored with even greater precision, to confirm the safety of people and property and to confirm that the behaviour of the dam is consistent with the design assumptions. In this respect, it is necessary to be aware that dams can and do fail before the reservoir is filled, due to the effects of weaknesses, overloaded by the self-weight of the dam. A serious risk is also that of landslides or rockslides triggered by excavation, for roads, conduits, tunnel portals, spillways or for the dam itself, for example if the excavation removes the natural buttressing supporting a slope or if the excavated slope(s) are too steep.

Seismic risk to dams and reservoirs and the derivation of seismic parameters to be used in design are discussed in Chapter 9 and will not be dealt with here.

References

[1] Doornkamp, J. C., Brunsden, D., Russell, J. R., Kulasinghe, A. N. and Gosschalk, E. M. (1982). A geomorphological approach to the assessment of reservoir slope stability and sedimentation. *14th ICOLD Congress, Rio de Janeiro*, pp. 163–174 (Q54–R11).
[2] White, R. (2001). *Evacuation of Sediments from Reservoirs, Appendix 1, Reservoir data*. Thomas Telford, London, pp. 151–162.
[3] BS 5930-89 Code of Practice for Site Investigation, 1981 (BSI, London).

[4] Houlsby, A. C. (1976). Routine Interpretation of the Lugeon Water Test. N. Ireland. *Quarterly Journal of Geology*, **9**, 303–313.
[5] Houlsby, A. C. (1976). Engineering of grout curtains to standards. *Journal of the Geotechnical Engineering Division of the American Society of Civil Engineers*, **9**, 953–970.
[6] Dunnicliff, C. J. D. (1988). *Geotechnical Instrumentation for Measuring Field Performance*. Wiley, Chichester.

5. Hydraulic studies

Introduction
The science of the conveyance of liquids forms the core of reservoir engineering practice and therefore it is fair to say that hydraulics is closely related to all the subject matter of reservoir engineering, albeit in some cases indirectly. Hydraulics is the medium by which reservoirs are designed and operate. These guidelines should bring the role, importance and use of hydraulics in reservoir engineering into focus.

Since this subject is so fundamental and all-embracing, its importance must be obvious but the importance of ensuring that the hydraulic concepts, studies and investigations are correct may not be so obvious without the appreciation of the consequences of getting even apparently minor elements wrong. Brief examples are:

- inadequate irrigation supplies reaching the crops and causing crop failures
- inadequate drainage of irrigated land causing water-logging, salinization and crop failures
- vibration and cavitation causing failure of hydraulic turbines, valves and gates, resulting in shortage of power supplies and failure of spillways
- inadequate discharge capacity of spillways (perhaps due to insufficiently streamlined flow) leading to overtopping and potential failure of dams and reservoirs
- surge pressures and water hammer without adequate control, leading to damage in tunnels and rupture of penstocks
- lack of aeration in high-velocity flow in filling tunnels and pipelines and in flood discharge causing explosive and damaging pressures.

The tools for making hydraulic investigations and designs are as follows.

(a) *Past experience* — often in the hands of experts and specialists but widely written up over a long period of time and continuously updated.

(b) *Analytical methods* — usually based on established principles but increasingly put within the practice of engineers through the application of computers. The methods in use have been extensively validated empirically and by experience. Scales are not limiting.
(c) *Computational models* — based on real and usually variable and changing situations. A sufficient and reliable database is required, with monitoring of the real situations to validate the models. Scales are not limiting.
(d) *Physical models* — usually built in a laboratory or in the open to represent a prototype. Scales are limiting: it may not be feasible to mobilize in the laboratory the discharge required to model the flow in a large river. It may not be feasible to represent the characteristics of fine sediment at the scale of the model. A choice of the relative importance of streamline and turbulent flow has to be made in the design of model. The cost and time consumed by physical modelling can be daunting. Nevertheless, hydraulic models produce an invaluable visual representation of full-scale conditions.

In many cases the results obtained by some or all of the methods (a) to (d) can and should be obtained and compared to help to establish confidence.

This chapter lays emphasis on the use of physical and computational models in solving problems in hydraulics. However, attention is drawn to guidance on the following subjects that involve hydraulics and are discussed elsewhere in these guidelines.

- The design and use of open spillways and gated crest spillways for dams, see Chapter 7.
- The design and use of low-level flood discharge outlets for dams, see Chapter 7.
- Energy dispersal, including spillways, stilling basins and valves, see Chapter 7.
- Water conduits for reservoirs, see Chapter 10.
- Hydraulic valves and gates (including avoidance of vibration), see Chapter 12.
- Water distribution systems for irrigation (including balancing storages), see Chapter 15.
- Control of the distribution of water, see Chapter 15.
- Increase in flood discharge capacity, see Chapter 17.

Past experience

Fluid mechanics, based on universally valid theorems of energy, momentum and continuity, form a logical part of the sciences of mechanics and physics. The problems which are not amenable to exact mathematical solution may still be accurately solved experimentally in accordance with the laws of similarity.

An appreciation of the scope of hydraulic engineering can be developed by reviewing the traditional types of water engineering projects such as multi-purpose dams, irrigation systems and navigation works and by taking note of recent developments in the field. By focusing on modern developments, the vastness of hydraulic engineering applications becomes apparent.

The need to be more efficient in the management of water has also required better techniques in hydraulic analysis. The hydrologic developments of the past century included the mathematical watershed model. The numerical modelling of runoff, coupled with the flood routing of flow in open channels, has made it possible to develop invaluable warnings of floods through both direct measurement of flow and precipitation and the on-line analysis of real-time data for quick flood predictions and public warning. The emphasis on large projects and the need to use energy efficiently have led to many technical advances that allow hydraulic engineers and hydrologists to analyse hydraulic and hydrologic systems with increased certainty.

Practically every problem in engineering hydraulics involves the prediction, by either analytical or experimental methods, of one or more characteristics of flow by three different methods. The first, and by far the oldest, is that of 'engineering experience' gained in the field by each individual engineer. The second, and relatively well developed, is the laboratory method of studying each specific problem by means of physical scale models and computational models. The third, neither new nor yet fully accepted, is the process of theoretical analysis. Each of these approaches has its obvious advantages and disadvantages. Field experience brings engineers closer to the problem in most practical aspects, but its development too often depends upon relatively wasteful methods; furthermore, the feel for the solution of field problems often lies within the person who has acquired experience. Model studies also display certain aspects but with a greatly reduced time scale and are generally restricted to the solution of specific problems. However, the results can be generalized to the extent that physical reasoning is involved in their analysis. The theoretical approach is, paradoxically, that which is most apt to provide either a useless or universal solution.

Theoreticians are farthest removed from the practical aspects of the problem, yet their process of reasoning, once the basic premises are established, is by far the most rigorous. The most effective solution of almost any problem can be obtained by combining the best features of all the above three methods of approach. The net result is intended to serve practical engineers as a firm foundation upon which to base their own analysis of problems yet to be encountered in reservoir engineering practice.

Analytical methods

The science of theoretical hydrodynamics was evolved from Euler's equations of motion for a frictionless (non-viscous) fluid and achieved a high degree of completeness. However, the results of classical hydrodynamics stood in glaring contradiction to experimental results particularly as regards problems encountered in water resources engineering practice. For this reason, practical engineers, prompted by the need to solve the important problems arising from the rapid progress in technology, developed their own highly empirical science of hydraulics. The science of hydraulics was based on a large amount of experimental data and differed greatly in its methods and in its objects from the science of theoretical hydrodynamics. A high degree of correlation between theory and experiment paved the way to the development of fluid mechanics in the nineteenth century.

It has been realized that the discrepancies between the results of classical hydrodynamics and experiment were, in many cases, due to the fact that the theory neglected fluid friction. Moreover, the complete equations of motion for flows with friction (the Navier–Stokes equations) had been known for a long time. However, owing to the formidable mathematical difficulties connected with the solution of these equations, the progress to a theoretical treatment of viscous fluid motion was slowed down. This difficulty was overcome during the early part of the twentieth century by Ludwig Prandtl who showed how to unify these two different branches of fluid dynamics. Prandtl achieved a high degree of correlation between classical hydrodynamics and experiment and paved the way to the remarkably successful development of real fluid flow.

The most commonly used important fluids in engineering practice, namely water and air, have very low viscosity and consequently the forces due to viscous friction are generally very small compared with other forces such as gravity and pressure. For this reason it is very difficult to comprehend that the friction forces omitted from classical hydrodynamic theory influenced the motion of fluid to so large an extent. In hydraulic investigation, the analytical methods are used in calculating the flow depth, rate of flow, velocity, pressure, etc., based on classical hydrodynamics equations. These methods involve the process of solving a number of partial differential equations governing the flow phenomenon. The basic equations of motion in hydrodynamics comprise three types which are generally known as elliptic, parabolic and hyperbolic equations.

Elliptic equations are Laplacian and describe a steady flow phenomena such as groundwater flow. There are no real characteristics for an elliptic equation and the solution can be obtained analytically using the method of singularities, analytical and complex functions, conformal and other transformations. Parabolic equations generally describe a flow in a transition from the steady state to another state due to turbulent conditions of the flow. Parabolic equations have no real root and the solution can also be obtained analytically. Hyperbolic equations describe an unsteady flow phenomenon including propagation of surges and waves in open channels and pipes. Hyperbolic equations have two real roots and it is not possible to obtain solutions analytically. However, it is a common practice to combine these three types of equation in order to obtain solutions through the method of characteristics, finite difference, finite element and boundary element methods. The method of characteristics can be carried out by manual graphical procedure or using a computer.

The finite difference method can also be carried out manually but it is more suited to solving equations by the use of digital computers than is the method of characteristics. In the finite difference method, the solutions are obtained by adopting explicit and implicit schemes. The explicit scheme progresses from one grid point to another using known boundary conditions to obtain the solution. The implicit scheme on the other hand utilizes recurrence relationships between unknown variables and interval or level of time adopted in the scheme, in order to march into the calculation with known boundary values at both ends of an actual model.

The finite element method is particularly suitable for approximate irregular flow boundaries. This method is most suited to the use of digital computers, as finite grid patterns can be generated to improve the accuracy of the solution of differential equations. The boundary element method uses known boundary values of the flow domain as a whole to solve the differential equations.

Computational models

With the advent of high-speed digital computers, computational fluid dynamics (CFD) has contributed enormously to computations in hydraulic engineering. Hydraulic studies involving complex processes of large-scale mathematical models were made possible to solve flow phenomena in rivers, estuaries and open channel networks. However, it is important that the model must be based on a sound basis of the physical flow phenomenon which requires correct theoretical equations of motion, sufficient field data (topographical and flow characteristics data) together with the correct specification of boundary conditions. All computational models need to be calibrated and validated to give reliable results. The accuracy of the results obtained from computational models depends heavily on the accuracy of the input data required to run the models. (A description of the conjunctive use of mathematical models and hydraulic survey to model hydrological conditions in the vast Irrawaddy River Delta in Myanmar is given in Chapter 1.)

There are several commercially available software packages on hydraulics, hydrology and environmental engineering. However, a thorough understanding of the equations used in developing the algorithms, the assumptions used in developing these packages and the limitations of the use of any commercially developed package is essential prior to using them in reservoir engineering practice. The objective of developing a computational model is to obtain water levels and discharges as a function of time at different sections along a reach of a river. The theoretical equations of conservation of mass and momentum (Saint Venant equations) are generally solved by using the finite difference method. One-dimensional (1D) and two-dimensional (2D) hydrodynamic models have been developed over the past twenty five years to solve a variety of problems associated with river flows in engineering practice. These models are particularly used for large-scale river systems incorporating different hydraulic structures, in order to study the characteristics of flood flows to give water levels along the river reaches with subcritical and supercritical flow conditions. However, in many of these commercial software packages, modelling bridge structures presented a difficult task in accurately assessing the afflux of flow near these structures because of their different shapes and sizes. In addition to commercially developed software packages, computational models developed by individual consulting firms and organizations are available to model river systems to study flow characteristics, sediment transport, groundwater flow and water quality in rivers, estuaries and other water bodies. Most of the hydrodynamic models developed predict both water levels and discharges along the river under steady and unsteady flow conditions. Some of the more noteworthy software packages most widely used in hydrological and hydraulic studies include HEC-1 and HEC-2 (see addendum to Chapter 6), DAMBRK in the USA and FLUCOMP, SALMON-F and ONDA (now replaced by ISIS FLOW), MIKE 11, MIKE 21, MIKE SHE, DAMBRKUK and FLOWMASTER(UK) in Europe and elsewhere. Most of these models are based on the conservation of mass, and full non-linear Saint Venant equations governing the flow phenomena in rivers and open channel networks.

Finance for procuring reliable data for long reaches of river systems plays an important role in a successful simulation of flow phenomena with computer models. The models can of course only be used where inherent mathematical features of the flow are well understood and the capabilities of the model are appropriate to the flow phenomenon under consideration. It should be noted that computational models suffer from a lack of the physical visual impact of the flow phenomena and also they cannot represent flow patterns at a smaller scale than their generated grid sizes and the developed algorithms of the equations of motion. However, there are several advantages in using the computational models, such as:

- large-scale river systems and open channel networks can be modelled
- simulated model flow phenomenon can be stored for later use
- they are fast and less costly to operate than physical models so that alternative designs may be tested with minimal cost
- they are flexible to change and can thus incorporate alterations in topographical and flow characteristics of the river system
- there are no scaling effects introduced in the models.

It is essential that computational models must reproduce the prototype flow conditions before they can be employed for engineering design. The purpose here is not to describe the complexities of different models but only to present a brief overview of computational models. Therefore, here we shall outline three examples of widely used commercially available software packages: ISIS FLOW and FLOW-MASTER(UK), developed in the UK, and MIKE 11, developed in Denmark.

ISIS FLOW

ISIS FLOW is a software package developed by H R Wallingford and Halcrow in the UK. It is used for modelling steady and unsteady flows in networks of open channels and flood plains. In addition, ISIS FLOW also contains units to represent a wide variety of hydraulic structures including several types of sluices and weirs, including jagged topped weirs and to model head losses through bridges, closed conduits and culverts with several standard shapes. Other units include reservoirs (e.g. to represent flood storage areas) and junctions. Free surface flow is represented by Saint Venant equations for unsteady flow in open channels. Different algorithms for different flow conditions were developed. Two methods are available for computation of steady flow problems, namely the direct method and the pseudo Time-Stopping Method. Muskingham and Muskingham–Cunge based flood routing methods are also provided.

In ISIS FLOW, the model boundaries are represented as either flow–time, stage–time or storage–flow (rating curve) relationships including specifying tide curves and hydrological boundaries. The usual way to create, edit, run and view results for ISIS FLOW is through the geographical user interface — ISIS Workbench under Microsoft Windows, which provides the means of accessing simulation module, Network Visualiser, Graphic Managers for time series, long sections, cross sections and $x-y$ plots, Forms Editor for entering model data and Tabular Processor.

FLOWMASTER(UK)

FLOWMASTER software is based on the initial computer program developed by British Hydromechanics Research Associates (BHRA) in the 1960s to deal with the steady and unsteady flow conditions in pipelines, particularly the problems of pressure surges in pipeline systems. The refined version of FLOWMASTER is a comprehensive computer program for the analysis of fluid flow in piping systems. It is a unique system that combines real engineering knowledge with the most advanced techniques in computer software. It is a highly interactive graphics-oriented design aid that enables engineers quickly and accurately to simulate fluid flow behaviour in complex piping systems. It provides a number of user-friendly facilities to create graphical representations of piping networks to analyse the flow through them under a range of operating conditions, to modify conditions or change network components and analyse the network under steady and unsteady flow conditions, as part of an iterative design process. It can be used in a number of ways:

- as a design tool to build up systems in stages and add to or modify the design with ease and speed
- as a design checking tool to predict possible problem areas in proposed installations
- as a means of investigating operational difficulties or failures in existing installations
- as a means of investigating possible extensions to the use of existing networks.

It is marketed by Flowmaster International Ltd (Milton Keynes, England).

MIKE 11

MIKE 11 is a software package developed at the Danish Hydraulic Institute (DHI) for the simulation of flows, sediment transport and water quality in estuaries, rivers, irrigation systems and similar water bodies. It is an integrated package, linking a number of independent modules through a structured menu system. The core of the MIKE 11 system consists of the Hydrodynamic (HD) module, which is capable of simulating unsteady flows in a network of open channels. The HD module is developed from Saint Venant equations as usual with different algorithms for different conditions of flow. The results of HD simulation comprise time series of water levels and discharges. Associated with the HD module are rainfall–runoff models (NAM and UHM) which may be used to generate inflows to the HD module. Advection–dispersion (AD) calculations, including cohesive sediment transport and water quality and non-cohesive sediment transport (NST) calculations, may be carried out using special modules that utilize the results of an HD computation.

In addition, the NST module can be run in tandem with the HD module as a morphological module. Statistical modules are also available to analyse the various result files. The analysis of raw and processed data can be carried out using the Hydrological Information Systems (HIS). HIS is designed to access MIKE 11 databases and result files and is also a stand alone module which can be used with any type of time series database. It includes facilities for the calculation of flow duration curves, double mass curves and also extensive rating curve analysis.

For flow simulation with these software packages, both upstream and downstream boundary conditions have to be specified at both ends of model branches that are not connected at a junction. The choice of boundary conditions depends on the physical situation being simulated and the availability of data. The typical upstream boundary conditions can be a constant discharge from a reservoir and/or a discharge hydrograph of a specific event. The typical downstream boundary conditions can be a constant water level in a large receiving water body or a time series of water levels (e.g. a tidal cycle) and a reliable rating curve (e.g. from a gauging station). If the resolution of the time-varying boundaries is greater than the time step used in the simulation, intermediate values in the boundary conditions have to be determined by linear interpolation.

In conclusion, computational packages provide a fast and inexpensive way of handling large-scale models which would be impossible with physical models.

Physical models

The prediction of any derived flow characteristics generally involves a mathematical or graphical expression of the relationship between two or more physical quantities describing the flow phenomenon. The quantities to be determined are known as the dependent variables, and those which govern its behaviour are known as independent variables. For an equation or plotted curve in a graph to be physically correct, obviously the expressed relationship between dependent and independent variables must be numerically correct. The relationship must also be dimensionally homogeneous. In other words, the independent variables must be so combined that every term of the relationship will have the same dimensions as that of the dependent variable. Any physical quantity appearing in such a relationship may be expressed in terms of units of measurement involving one of three, using the four standard dimensional categories of mechanics, namely mass, force, length and time. Those generally used are mass or force, length and time ([MLT] or [FLT]) systems. The classes of dimensional variables involved in the mathematical or graphical description of a given state of fluid motion are the variables describing:

- the characteristics of flow such as the depth of flow (d) or (l), mean velocity (v) or rate of flow (Q) in a channel and the pressure distribution (p) around or the total force upon a body (e.g. a bridge pier)
- the properties of the moving fluid such as the density (ρ), absolute viscosity (μ), compressibility (K), and surface tension (σ)
- the boundary geometry such as height of land, head over a weir (H) or form and cross sectional area of flow passage (A).

In general, the dependent variable in any flow relationship is one of the characteristics of the flow itself, since the quantities which may be varied independently and which hence define the problem are normally restricted to those describing the boundary form and properties of the fluid, with the exception of one arbitrary flow characteristic. In order to check the validity of equations experimentally, it would be necessary to conduct a vast series of experiments in which each independent variable is made to change systematically over its entire physical range. This procedure would be very expensive and time consuming in obtaining a solution to the problem. Therefore, dimensional reasoning is a powerful initial tool in the analysis of a problem and in the process of combining variables to yield significant dimensionless parameters which govern the flow.

Basically, dimensional analysis is a method of reducing the number and complexity of experimental variables which affect a given physical phenomenon, using a combining technique with the help of an [MLT] or [FLT] system. As Newton's second law of motion provides a relationship between force, mass, length and time, the [FLT] system presents no conceptual problems. Although the purpose of dimensional analysis is to reduce the number of variables by grouping them in dimensionless form, it has several benefits in engineering practice.

The first benefit is the reduction in cost of an experimental investigation due to the grouping of variables. A second benefit of dimensional analysis is that it helps thinking and planning for an experiment or theory. It suggests variables which can be discarded and also gives a great deal of insight into the forms of the physical relationship of the problem being studied. A third benefit is that dimensional analysis provides scaling laws that can convert data of results from an inexpensive, small model into design information for an expensive, large prototype project. When a scaling law is valid, a condition of similarity exists. Although dimensional analysis has a firm physical and mathematical foundation, considerable art and skill are needed to use it very effectively; such skill requires time, practice and maturity to master.

Several methods for obtaining the dimensionless similarity parameters have been developed. These methods include Rayleigh's method or the method of indices, Buckingham's method or the Π theorem, the method of similitude and the systematic use of differential equations. Rayleigh's and Buckingham's methods are the least powerful and they are normally applied with ease. Systematic use of differential equations governing the flow phenomenon provides the greatest power for finding the solution, but it is the most complex to apply. If differential equations are available in reasonably detailed form, the systematic use of these equations should provide a solution superior to the other methods. Mathematical models having complete differential equations can be important as a primary source for obtaining a solution to a problem.

Rayleigh's method

Rayleigh's method essentially consists of writing down the function equation which formulates and defines the problem, rewriting this equation in terms of the dimensions involved and then equating the exponents of [M], [L] and [T] to ensure that the equation is dimensionally homogeneous. When formulated correctly, the dimensionless parameters can be used for the purpose of experimental physical model investigation.

Buckingham's method

Buckingham's method (or Π theorem) is a formal means of arranging dimensional variables of a physical equation in non-dimensional groups. All the variables can be organized into the smallest number of dimensionless groups by Buckingham's method. Although the method offers a better means of understanding the problem, it has certain limitations. In applying the Π theorem, it may be found that the theorem does not disclose conditions for neglecting Π terms. However, in simple problems, the theorem gives remarkably complete and accurate answers; in complex problems, it shows the usefulness of dimensionless groups. The method provides a basis for the analysis of units, dimensions, and relationships in a variety of engineering problems. Buckingham's method is somewhat similar to Rayleigh's method except that each Π term is derived independently from the others. The procedure is to select three relevant variables involving all dimensions present and to combine each other variable in turn with these three so as to form a dimensionless parameter. The method provides a remarkably reliable procedure and when the variables are properly chosen, dimensionless parameters often will correlate and generalize the results of an investigation.

Method of similitude

Dimensional analysis is directed primarily towards experimentation, the model laws being obtained by inspection of the dimensionless equations, whereas similitude analysis and similarity theory expressions concentrate primarily on developing model laws of similarity from which a dimensionless functional equation may be derived. Geometric similarity requires that length ratios between model and prototype are the same. Kinematic similarity is a correspondence of motion: in two kinematically similar systems, particle motion will be similar. Dynamic similarity occurs when the ratios of forces are the same in the two systems.

In applying the method of similitude to the design of models in hydraulic engineering studies, forces of importance, including the dependent and independent forces such as inertia ($\rho v^2 l^2$), pressure (pl^2), gravity ($\rho g l^3$), viscosity ($\mu v l$), surface tension (σl) and elastic compression ($K l^2$) are taken into consideration by the investigator. Correlating groups of these forces expresses the required similarity of two systems. Two systems having geometric, kinematic and dynamic similarity have complete similitude. The correspondence between a hydraulic model and its prototype is limited because the similitude for one or more forces is usually incomplete as no model fluid has the viscosity, surface tension and elastic characteristics to satisfy the conditions of complete similitude. Nevertheless, application of complete similitude is not difficult because, in the majority of hydraulic model studies, neglecting the effects of surface tension and elastic forces produces only minor scale errors which can be minimized by carefully selecting the model scale and the fluid. The scale selection should be made to satisfy the predominant force and neglect the remaining forces. Most models can closely simulate fluid motion if either gravity or viscous forces predominate; correction for the effects of other forces is required only occasionally.

Laws of similarities

Laws of similarities for hydraulic model studies can be derived from dimensionless parameters for each of the above forces related to the inertia force as illustrated below:

Euler number	= inertia force/pressure force:	$\rho v^2/p$; or $v/(p/\rho)^{1/2}$
Froude number	= (inertia force/gravitational force)$^{1/2}$: (*l* being characteristic dimension such as depth of flow)	$v/(gl)^{1/2}$
Reynolds number	= inertia force/viscous force: (where μ is the absolute viscosity of the fluid)	$\rho v l/\mu$
Weber number	= (inertia force/surface tension force)$^{1/2}$: (where σ is the surface tension of the fluid)	$v/(\sigma/l)^{1/2}$
Mach number	= (inertia force/elastic force)$^{1/2}$: (where K is the bulk modulus or compressibility of the fluid).	$v/(K/\rho)^{1/2}$

The above non-dimensional numbers give corresponding laws of similarities based on the predominant forces present in the investigation of model studies. In many engineering problems it is not economical to make full-scale trials, so the experiments on small-scale models based on the above laws of similarities are invaluable. In most hydraulic model investigations only gravity and viscous forces are ordinarily of importance and hence the models are designed based on the corresponding velocity-scale ratios in each case. It is important to note that, although pressure forces are very important, the law of similarity based on Euler number is not essential, because it reflects the inverse of the coefficient of drag (C_d) in the general drag-free equation $F = C_d(\frac{1}{2}\rho v^2 l^2)$ used in hydraulic model studies to estimate the forces on the prototype structures.

Differential equations

These equations describe the physical and mathematical conditions of the problem and therefore provide a basic element in the preparation and use of computational models described in the previous paragraphs. Differential equations written for an elemental volume of fluid are the most complete and detailed equations governing the fluid motion. This description becomes a mathematical model limited to what are considered to be the important physical aspects of the flow phenomenon. Those aspects excluded determine the acceptance of uncertainty of the model and the inaccuracies acceptable for given circumstances. The general procedure is that the differential equation containing dependent and independent dimensional variables is made dimensionless through the process of normalization. The usefulness of the dimensionless parameters depends on the care and insight applied in choosing the forms of important parameters. In applying this method to model studies, the investigator must understand the physical aspects and the limitations of mathematical models in spite of the equations being carefully defined in form and magnitude by the dimensionless parameters. Boundary conditions have to be established for all dimensionless parameters present in the equations. Much has been covered already on mathematical and computational models in the previous sections, therefore the following sections are devoted to the use of physical models in hydraulic engineering practice.

Application of similarity laws to physical models

In applying the similarity laws to physical models, satisfying more than one law of similarity in a model investigation requires that the physical properties of the testing fluid be variable over rather broad limits. For example, to satisfy the Froude and Reynolds laws of similarity simultaneously, it would be necessary that the kinematic viscosity ratio between the model and the prototype is related to the length scale ratio to the power 1·5. Satisfying this criterion for any hydraulic model with a length scale ratio of 1 to 25 would require a test fluid having kinematic viscosity 1/125 that of water. Such a fluid is not available. The physical properties of a practical testing fluid are such that only dominant parameters can be modelled based on the predominant force in a given situation.

Froude law of similarity

The Froude parameter is used most frequently in hydraulic engineering problems involving turbulent free-surface flow conditions. Conditions assumed in formulating the parameter are essentially realized in the case of turbulent flow with a free water surface because the effects of gravity outweigh those of viscosity and surface tension. However, when viscous and surface tension forces are neglected, every effort should be made to minimize them by using large-scale models with smooth boundaries. Generally, when the Reynolds number of the model is greater than 1×10^4 with the depth of flow used as the characteristic length in defining the Reynolds number, the viscous forces are relatively unimportant. Compensation adjustments to slope or boundary roughness can be made if viscous effects cannot be ignored.

Reynolds law of similarity

Steady flow in pipes, or flow around a deeply submerged body, approximates the conditions assumed in formulating the Reynolds law of similarity. With no free surface directly involved in the flow pattern, and with steady flow, the forces of surface tension and elasticity are eliminated. Furthermore, the gravity forces are balanced and therefore do not affect the flow pattern. The Reynolds number in the model can rarely be equal to that of the prototype exactly, but fortunately this is not absolutely necessary. The reason for this is that for all values of relative roughness less than 0·001, the friction factor becomes nearly constant when the Reynolds number is greater than 1×10^6 and the resistance forces, model to prototype, depend only on the ratio of relative roughness. In order to obtain a satisfactorily large Reynolds number in a small model, the velocity is often increased by using larger than scaled pressure heads to provide similar boundary resistance forces, thus resulting in Reynolds number in the model being not exactly equal to that of the prototype.

Design, construction and operation of hydraulic models

Prototype information for use in designing a model must include where relevant, topographical maps, drawings depicting an overall plan and cross sections of the proposed or existing prototype project. Sufficient detail must be available to determine the shapes and characteristics of all surfaces over which the flow passes. Hydraulic details of other related structures may be necessary. Knowledge of the types and condition of materials composing the river bed and their boundaries is needed to establish the roughness of the model channel. A reliable water stage–discharge curve for the river downstream from the proposed structure is required. In analysing river and estuary problems, historical and predicted data, such as discharge, water stage, tide, salinity, sediment carrying capacity and depths of flow have to be obtained. For spillways and other outlet works, proposed operating conditions and the range of discharges to be expected have to be specified or otherwise determined.

Success in achieving the desired results from a laboratory study in the least time and with least expense depends largely on the design of the model. A highly important aspect in the design of model is the selection of a model scale. Generally, scales as large as feasible should be adopted, as permitted by finance and the facilities of available space and electricity and water supply in the laboratory. The following scale ratios have been successfully used in model studies and may provide a guide:

- spillways of large dams 1:30 and 1:100
- medium size spillways 1:60
- outlets — gates and valves 1:5 to 1:30
- canal structures — chutes and drops 1:3 to 1:20

- river model horizontal scales 1:100 to 1:1000
- river model vertical scales (distorted model) 1:20 to 1:100

Models of valves, gates and conduits should have flow passages at least 100 mm to 150 mm across so that, for the heads and discharges normally available in the laboratory, turbulent flow can be produced in these appurtenances. The scale of an outlet model is often determined from the minimum diameter or width. In models of canal structures, the bottom width of channels should exceed 100 mm so as to allow turbulent flow in the model.

To reduce the effects of viscosity and surface tension, spillway models should be scaled to provide flow depths over the crest of at least 75 mm for the design of the model in the normal operating range. Time and expense may be saved by choosing the scale for tunnels or conduit models to accommodate a standard pipe size or sections of pipe available. Parts of previously tested models can be used to cut costs.

If the surfaces over which the water flows are reproduced in shape, and the roughness of the surfaces is approximately to scale, the model will usually be satisfactory. Initial construction of the model should allow for considerable modification with minimum costs in rebuilding it. Flow visualization in various parts of a model may be very important and this may warrant the use of transparent plastic materials in the construction of models to facilitate photographic recording of flow patterns.

The operation of a model should be carefully planned to evaluate the success of the design study. In general, the evaluation includes proving the model behaves like the prototype by qualitative and quantitative tests to meet the operational requirements. Most models need to be adjusted and retested. The adjustment phase includes preliminary trials to reveal model defects and inadequacies. Testing of the model should include a systematic examination of each feature of a proposed design for improvement in its operation and reduction in construction and maintenance costs. As the study progresses, functional relationships among different variables should be examined to aid in detection of measurement errors. The model results thus obtained must be correctly interpreted and generalized so that the information can be used for verification studies by actual field measurements. The end product of any model study is the report, which transmits the findings and recommendations. It is important to note that a complete photographic record of all important tests is indispensable and often helps to eliminate the necessity of repeating tests.

Free-surface flow models

Flow over hydraulic structures such as spillways and weirs involves significant vertical components of velocity and it is therefore necessary to construct the model to natural or undistorted scale. Any exaggeration of the vertical scale would cause an unacceptable increase in vertical components and therefore any such vertical distortion must be avoided. As the flow is mainly dominated by gravitational effects, models of free-surface flow around hydraulic structures must be scaled according to the Froude law of similarity. To ensure that the model is free from scale effects, the model must be fairly large and scale ratios ranging from 1:30 and 1:100 are recommended. In achieving geometric similarity, it is not necessary to attempt to scale the roughness of the prototype which is usually small so that the model should be as smooth as possible. In some cases it may be necessary to construct a composite model of a scheme in order to predict the performance of the prototype project.

Fixed-bed models

Fixed-bed models are used for canals and rivers having relatively long stretches of stable bed configuration. Studies with these types of models include backwater effects caused by obstructions or channel improvements, flood routing and determination of flow distribution in estuary channels and river flows. Special attention must be given to the model boundary resistance so that the Froude law of similarity is truly applied. For large models, the velocity may be high enough to give turbulent flow and the minimum model roughness may have smooth enough boundaries to represent the prototype properly, while small models may not yield satisfactory results because boundary roughness prevents a scaled velocity. Therefore small models require a distortion in vertical scale and/or slope to offset the disproportionately high resistance of the model boundaries and to obtain a sufficiently high value of Reynolds number to ensure turbulent flow in the model. The distortion in slope can be computed with sufficient accuracy by using Manning's equation. The model boundaries may also have to be made rougher by artificial means to compensate for the exaggerated slope. Generally, a model with roughness adjusted for a particular depth of flow will yield dependable results for flow at or near this depth. However, when the model is run with several depths of flow, the model roughness should be adjusted to give an average resistance that is approximately correct over the desired range, or the roughness may be varied with depth for reasonable accuracy. One of the methods of varying roughness is to place sections of wire screen in the channel to simulate the correct roughness. The fact that should be remembered is that Manning's n is strictly a roughness characteristic and that Manning's equation applies only at sufficiently high values of Reynolds number where resistance forces are proportional to the square of the velocity. The most important precaution in testing distorted river models is to make sure that the flow in the model is of the same type as in the prototype.

Movable-bed models

Fully comprehensive studies of open channel problems involving scour, deposit and transport of sediment in rivers and streams require movable-bed models. Despite the limitations and roughness of modelling materials, movable-bed models prove to be very valuable in solving complex problems involving the transport of streambed materials. Mathematical analysis for similitude in movable-bed models is not as rigorous as that applied to models of hydraulic structures having fixed boundaries. A successful study with movable-bed models requires a balance of hydraulic forces to produce bed movement having the same character observed in the prototype.

The general design of the model requires selecting scales and materials that will result in bed movement similar to that in the prototype for the estimated range of discharge. The usual model scales result in sediment particles that are so small they no longer act like bed material; instead, the material tends to become either suspended in the flow or compacted into an unyielding bed. Therefore, bed material of larger particle sizes is used to offset the scaling effect. Geometrical distortion of the model becomes necessary to produce velocities that are sufficient to move the bed material. In order to obtain these conditions, distorted models are usually preferred, with exaggerated slopes to produce sufficient tractive force. The distortion, usually defined as the vertical scale ratio divided by the horizontal scale ratio, is kept in the range of 2 to 9, but generally, distortion should be aimed at as low a value as possible without appreciably reducing bed movement. The difficulties of reaching an acceptable balance between boundary resistance and bed movement can be minimized by using model bed materials (e.g. coal, sawdust, pumice, plastics, etc.) with lower specific gravity than the prototype bed material.

The verification of a movable-bed model is an intricate process. The accuracy of the functioning of the model must be established and certain scale ratios, such as time and discharge, have to be determined experimentally. Therefore, the verification stage of a movable-bed river model study and the interpretation of results are largely dependent on general judgement and reasoning.

Comparison of models

There are several similarities between the investigation requirements of physical and computational models. They all need topographical data of the prototype and they need to be calibrated and verified. The models should also simulate the prototype conditions of flow. In a physical model, the river topography is reproduced to scale whereas in a computational model the river topography is represented by a series of co-ordinates for each cross section along the reaches of the channel network. The roughness in a physical model is adjusted to represent the prototype conditions, whereas a computational model is calibrated by altering the value of roughness coefficients and co-efficients of velocity heads for each cross section until the predicted and observed water levels and discharges in the prototype are reproduced. The design features are specified in a physical model by flow rates and water levels, while they are specified by values for the inflows and water levels from the databases of a computational model. The changes in river bed topography are moulded to represent new geometrical features and the same is achieved by altering the co-ordinates in a computer model. The scale effects must be minimized in a physical model while there are no scaling effects except some minor numerical errors in a computational model. However, it is essential that both types of model reproduce prototype conditions before they can be utilized for design purposes.

In conclusion it is reasonable to state that, in spite of the advances in the development of high-speed digital computers to deal with the modelling of turbulence, physical hydraulic models will always play a significant role in civil engineering design practice.

Bibliography

[1] American Society of Civil Engineers (eds.) (1989). *Civil Engineering Guidelines for Planning and Designing Hydro-Electric Developments.* Energy Division, American Society of Civil Engineers, New York.
[2] Creager, W. P. and Justin, J. D. (1950). *Hydro-electric Handbook.* Wiley, Chichester. 2nd edition.
[3] Davis, C. V. and Sorenson K. E., (1956). *Handbook of Applied Hydraulics.* McGraw-Hill, New York. 3rd edition.
[4] Francis, J. R. D. and Minton, P. (1984). *Civil Engineering Hydraulics.* Edward Arnold, London, 5th edition.
[5] Guthrie Brown, J. (Ed.). *Hydroelectric Engineering Practice.* Blackie, London. (3 vol. work in 2 editions.)
[6] The Institution of Civil Engineers (1996). *Floods and Reservoir Safety.* Thomas Telford, London, 1996, 3rd edition, pp. i–viii, 1–63.
[7] Morris, G. L. and Fan, J. (1998). *Reservoir Sedimentation Handbook — Design and Management of Dams, Reservoirs and Watersheds for Sustainable Use.* McGraw-Hill, New York.
[8] Rouse H. (1938). *Fluid Mechanics for Hydraulic Engineers.* McGraw-Hill, New York.
[9] Ven Te Chow. (1981). *Open Channel Hydraulics.* McGraw-Hill, New York.
[10] Warwick, C. C. (1984). *Hydro Power Engineering.* Prentice Hall, London.

[11] Webber, N. B. (1971). *Fluid Mechanics for Civil Engineers*. Chapman and Hall, London.
[12] Wright, D. E. (1971). Hydraulic design of unlined and lined invert rock tunnels. CIRIA Report 29. London.
[13] Allen, J. (1947). *Scale Models in Hydraulic Engineering*. Longmans, Green, London.
[14] American Society of Civil Engineers (1942). *Hydraulic Models, ASCE Manual of Engineering Practice*, No. 25. ASCE, New York.
[15] Henderson, F. M. (1966). *Open Channel Flow*. Macmillan, New York.
[16] Novak, P. and Cabelka, J. (1981). *Models in Hydraulic Engineering — Physical Principles and Design Application*, Pitman, London.
[17] Yalin, M. S. (1971). *Theory of Hydraulic Models*. Macmillan, London.
[18] Cunge, J. A. (1975). *Two Dimensional Modelling of Flood Plains in Unsteady Flow in Open Channels*. Edited by K. Mahmood and V. Yevjevich, Water Resources Publications, Colorado.
[19] Cunge, J. A. et al. (1980). *Practical Aspects of Computational River Hydraulics*. Pitman, London.
[20] Fread, D. L. (1988). *Dambrk, the NWS dam break flood forecasting model*. Office of Hydrology, National Water Service (NWS), Silver Spring, Maryland, 20910, USA.
[21] Price, R. K. and Samuels, P. G. (1980). A computational hydraulic model for rivers. *Proceedings of the Institution of Civil Engineers*, Pt. 2, Vol. 69, March 1980, pp. 87–96.
[22] Samuels, P. G. and Gray, M. P. (1982). The FLUCOMP river model package, an engineer's guide. Report EX 999, Hydraulics Research Ltd, Wallingford, Oxon, UK.

6. Hydrological studies

Background

Generally, hydrological studies relating to reservoirs are concerned with yield assessment and the estimation of the flood which the reservoir must be required to convey safely. Yield is the volume of controlled water released from a reservoir during a given time interval and is an indicator of the capacity of the reservoir to release regulated flow over that time interval [1].

The subject of hydrology was introduced in Chapter 3, by summarizing the data and the basic processing of the data required. Cyclical influences and the possibility of long-term climatic change were mentioned. There followed a description of the preparation of flow duration curves and mass curves for rivers, in a preliminary way. The evaluation of the hydrology of a river basin is the basic premise on which is founded the sizing of reservoirs in that basin and the output of the projects which they serve. It is therefore all important to obtain a sound and realistic evaluation and this must be undertaken by a suitably experienced hydrologist to confirm the hydrological viability of the scheme. The objectives are no less than the correct sizing of dams and reservoirs, spillways, conduits and the development to be served. It is not the intention here to provide an exposition of the detailed science or methods of hydrology but to convey the knowledge of hydrology required for a basic understanding of the evaluation of water resources for reservoirs.

The catchment of a river basin or reservoir is broadly defined by the line connecting the highest points around the whole basin or reservoir above its dam site. On one side of the line, surface water runoff from rainfall will flow towards the river or reservoir, while on the other side it will flow away and be lost to other rivers, lakes or the sea. One must not lose sight, however, of the potential for some of the rainfall on the catchment to be lost by underground flow to the valley downstream of a reservoir, or to rivers, lakes or the sea outside the catchment area. Underground flow and underground storage in aquifers may increase as well as diminish the potential water resources of the catchment area. Thus the hydrological evaluation of the water resources available involves 'hydrogeology'. It is also important to account for the influence on catchment boundaries of embankments constructed for roads, canals, flood defences and railways, etc., in other words, artificial modification of the natural topography.

The possible contributions to evaluation by geomorphologists and geologists of erosion and sediment transport in tributaries and rivers was indicated in Chapter 4; specialists in hydrogeology may be required to assess underground storages and flows; specialists in sedimentology are trained to undertake the overall study and to estimate the rates, timing and distribution of sediment deposition in a reservoir (see Morris and Jiahua Fan, 1997 [2]). Initially, however, at least in the earlier studies, it is the hydrologist who is usually expected to undertake an overall preliminary study, in consultation with the geologist. Other specialists may be needed later, depending on the nature and significance of the problems as assessed by the hydrologist.

The outcome of the studies should be an as-refined-as-possible development of the initial mass curves for reservoir inflow and outflow, described in Chapter 3. These can be integrated (from a plot or tabulation of reservoir volume against reservoir level), on a time basis, with increments of reservoir volume and reservoir level, rising or falling, to produce a simulation model of reservoir operation, using changes in reservoir storage to give changes in reservoir level. The refined curves for simulation of reservoir operation should have input data with a 10-day or, if practicable, daily module, allowing for losses due to leakage and evaporation from the reservoir. Simulated future discharges will be varied to suit predicted demand, which is variable in accordance with use and climate, as outlined in Chapter 2. In the case of irrigation projects, for example, it will be varied primarily according to the crops to be grown, crop seasons, rainfall on the irrigation area, water available from other sources (e.g. from other rivers or tributaries) and soil retention. The methods used to determine inputs to the model can be either deterministic or probabilistic and are often a combination of both.

Deterministic methods
The hydrological evaluation can be undertaken from first principles on a deterministic basis: the determination of river flow from rainfall — 'the rainfall–runoff method'. The rainfall precipitation is routed from its point of fall, along the route it would follow (usually the steepest) to the river by open channel flow theory, allowing for friction (roughness) and surface gradient along the route, both of which may vary.

Initially the flow is over the surface to a tributary, and then by tributary channel to another tributary or the main river. Losses are dependent on the absorptive capacity of the surface, evaporation, and the time taken (velocity of flow). The result of the calculation should be a hydrograph of flow in the river, showing the build-up and decline of flow in the river with time, during and after periods of rainfall. Alternatively, runoff may be calculated by the application of a 'runoff factor' (k) to rainfall, measured as the total rainfall on an area in 24 hours, in one month or in one year. The factor, k, takes account of all losses and is usually derived from standard values published, or from back-analysis of measured runoff and rainfall on the same or similar catchments. Clearly, k is subject to the topography, length and gradient of the valley slopes, surface cover and absorption by the ground. It will vary with elevation and wind direction in relation to the slopes. The main catchment should be divided into sub-catchments, including the catchments of separate tributaries with different characteristics. If the catchment is subject to sub-zero temperatures and snow fall, there will be a lag in release of runoff dependent on snow-melt, which will occur at rates depending on air temperatures and hours of sunshine. With a sudden occurrence of warm weather, snow-melt can be very rapid, causing intense floods, which may be magnified by concurrent rainfall and by discharge from springs.

A period of record of river flow can be created, infilled or extended synthetically (synthesized) by correlation of the measured flows with rainfall, for which the records usually exist for a longer period than the records of river flow. It must be accepted that it is surprisingly difficult to establish correlations of rainfall and runoff, with a good significance. Extended records may also be synthesized (but less reliably) by correlation with records for neighbouring rivers and/or with the cyclical influences, mentioned in Chapter 3.

The extended time series of records and correlated values can then be fed into the reservoir operation simulation model outlined. The output of the model will show the reservoir storage used to fulfil the demand and will depict the critical periods of drought when the reservoir is drawn down under extreme conditions, to minimum water level or even below. It will also show the periods of high flows and floods when water would be spilled, in effect, to waste (as far as the reservoir in question is concerned). The length and timing of the critical periods should be noted and the time series of rainfall and river flow should be examined on a daily basis or even an hourly basis, to see if more severe sequences occurred which might have taken place at the critical times for the model, with resulting more severe shortages in output than those revealed by the time module adopted in the model: averaging conceals short-term peaks and troughs. The model must allow for any minimum river flows or compensation water discharges which the project is required to make up and sustain by releases from the reservoir if necessary, for the benefit of riparian/riverine inhabitants in the valley below the reservoir. In general, it is usual to expect that reservoirs will be operated during floods, so that discharges do not exceed the river flows which would have occurred in that length of river, had the reservoir not been built. This condition is intended clearly to demonstrate that flooding below the dam is never worse than it would have been naturally. It is also important to ensure that the rate of opening of hydraulic gates does not cause a worse build-up of discharge and possibly more severe surge waves than would have occurred naturally, since such unexpected waves can wash away people and property in or near the river at the time. Thus the operational plan for spillway gates needs to be tested theoretically under the flood hydrographs adopted for design, for both the with- and without-reservoir conditions.

The average regulated discharge of the reservoir (the long-term yield) can be obtained directly from the output of the reservoir simulation model (Chapter 3), as can the additional 'unregulated' discharge by overspill during high flows (which may have value in providing temporary increases in output and perhaps in saving in outputs which would have to be provided from other sources).

Flood studies

It should be appreciated that (in general) peak flood discharge per unit of catchment area logically increases as catchment area decreases. This is firstly because the intense rainfall bursts which cause extreme runoff are usually of short duration and occur over a relatively small area. As catchment area increases, their effect in generating extreme flood runoff is averaged out and moderated. Secondly, the influence of storage of flood water, both on the catchment and in the channel system, attenuates (decreases) the peak rate of flow per unit area of catchment for a given depth of storm rainfall. This effect clearly increases with catchment area.

Similarly, for a given region, the ratio of the flood with the recurrence interval T years in discharge per unit area (Q_T), to the mean annual flood in the same units (Q_{ave}), is proportional to $\log T$. Typically (in temperate regions) the ratio of the 1:100 year maximum flood discharge to the annual average would be around 3 to 3.5 for a 200 km^2 area, decreasing to 2 to 2.5 for areas of 500 km^2 or more. This again reflects the fact that extremes (in terms of storm rainfall depth and therefore flood runoff per unit area) tend to moderate as area increases.

The maximum floods which the reservoir should be designed to discharge, store or spill can be estimated by mainly deterministic methods or mainly probabilistic methods — or both. However, the application of past records of rainfall or river flow to postulate 'return periods' or 'frequency' of rainfalls or river flows effectively acknowledges the probabilistic nature of the resulting values, which is implicit in most estimates, which necessarily involve statistical methods. This section is concerned with the estimation of flood peaks by risk-based and probable maximum methods. Probable maximum methods point to the probable maximum flood (PMF) or the probable maximum precipitation (PMP) (which leads to the PMF). Probabilistic estimates are enveloped by, but not necessarily asymptotic to, probable maximum estimates. Methods of linkage of flood frequency curve to PMF and associating a return period with PMF are referred to in Chapter 7. Flood volume can be associated with estimates of flood peak through methods which model the transformation of rainfall to runoff.

Where there is lack of data, there has to be resort to the more generalized regional methods, while the more specific and advanced implementations are practised in the UK and elsewhere where records, technological development and environment permit.

It is noteworthy that an eight-volume report *Australian Rainfall and Runoff* [3] was produced by the Institution of Engineers, Australia in 1999 from which wide-ranging guidance can be obtained on flood plain management, detention basins, river works, urban trunk drainage and so forth.

Methods for flood estimation in the UK subject to validation October 2000

The *Flood Estimation Handbook* (FEH) [4] was published in 1999 as a result of research from 1994 to 1999 funded by the appropriate authorities concerned in the UK. This handbook aimed to provide clear guidance to practitioners concerned with rainfall and flood frequency estimation in the UK and intended to be more nearly a handbook of estimation methods than a report of flood studies. It comprises five volumes:

1. Overview
2. Rainfall frequency estimation
3. Statistical procedures for flood frequency estimation
4. Restatement and application of the *Flood Studies Report* (FSR) rainfall runoff method
5. Catchment descriptors.

The FEH presents types of analysis (deterministic and probabilistic) mentioned at the start of this section but does not deal with envelope methods. Volume 2 of the FEH presented a new generalised rainfall frequency estimation procedure for the UK. However, substantial inconsistencies were discovered between results obtained from applying the FEH methodology and that of its predecessor, the *Flood Studies Report* (FSR) [5], published in 1975 in collaboration with the Natural Environment Research Council (NERC). On publication, it was stated that the FEH superseded most of the FSR, the use of which had become cumbersome, chiefly as a result of complicating revisions. The FEH adapted the FSR rainfall–runoff method for use with the FEH catchment descriptors and replaced the FSR rainfall frequency model with a wholly revised rainfall frequency model. The FSR PMP methodology remained the only generalized procedure for estimating PMP available for the UK while, in accordance with its title given above, the FEH Volume 4 is a restatement and application of the FSR rainfall–runoff method, of which the heart is the unit hydrograph and losses model. From the differences found between the values of floods estimated by the FEH and the FSR, it was postulated (in illustration of the differences) that either:

- the FEH methodology provides high estimates for events with a 10 000 year return period; or
- the 10 000 year and PMP estimates from the FSR methodology are too low; or
- both the FEH and the FSR methodologies are valid and the range of estimates truly reflects the uncertainty of those estimates.

It was clear that the implications of the anomalies for reservoir safety (and flood defences, etc.) were far reaching and the Department of the Environment, Transport and the Regions (DETR) commissioned a study on behalf of England, Scotland and Wales, to provide further advice. The result was a report by consultants to the DETR [6]. The report concluded that the FEH method was yielding rainfall estimates seriously greater than those currently used in reservoir flood assessments.

A background and departmental response to this report [7] was produced by the DETR, the final draft comprising nine pages dated October 11th 2000. The DETR response indicated that the scope and timetable for the recommended validation exercise and research would be confirmed to Centre for Ecology and Hydrology (CEH) Wallingford before the end of 2000, while longer term thinking regarding PMP and reservoir safety was continuing. The Department invited the Institution of Civil Engineers to establish a group to steer the recommended Projects and, in due course, also revision of *Floods and Reservoir Safety* [8]. The Department endorsed interim guidance for Panel Engineers suggested in the Babtie Group/CEH report of September 2000, on how to assess reservoir flood safety, and invited comments within 12 working weeks.

The Author is impelled to say that the inconsistencies found between the FEH and the FSR were very disconcerting. While conservative, cost-effective interim measures can be taken, it appears to the Author that it may take the order of perhaps five years to achieve the high level of confidence desired. It appears to be believed by those closely concerned that some of the reasons for the discrepancies found are understood (for example, inadequate statistical suitability of the location of gauges with sufficient records) but the resolution of anomalies may be very time consuming. The questions raised include design standards for floods and reservoir safety, some of which are likely to be controversial, with considerable implications for cost in dam construction. In passing, the Author's view is that more fundamental error(s) in the methodologies than initially appreciated may be discovered. It would be inappropriate for the Author to try to pre-empt the findings which will emerge from the investigations which are being urgently undertaken under the authority of the DETR (now Department of Transport, Local Government and the Regions) and, if possible, readers of these guidelines may best be advised to await the issue of conclusions from the DETR, the British Dam Society and the British Hydrological Society. However, while both the methods of the FSR and the FEH are in question, rather than omit reference to either, the Author thinks that in many respects both will prove of value in future and includes an introduction and outline of both here.

Flood studies: first deterministic method

From the sequence of rainfall in the time series which would produce the most critical combination of volume and peak intensity of inflow into the reservoir when estimated by the 'unit hydrograph' method [9].

This is simply a development of the deterministic method which has been outlined, of deriving hydrographs from runoff, applied to sub-catchments of the river above the dam. The greatest volume of inflow will arise from a storm covering the whole catchment, so the inflows from each sub-catchment are cumulatively combined in keeping with their times of concentration and the predicted rainfall distribution of the storm. The storm will not have uniform rainfall intensity except over very small catchments, since the rainfall intensity tends to attenuate outwards from the point of highest intensity. The highest intensity of discharge at a point on the river, e.g. the dam site, is usually not reached until a 'time of concentration' has elapsed, sufficient for discharge from the location of highest rainfall to reach the dam site, before it and other contributing runoffs start to attenuate. It is unusual for storms lasting for periods of less than the time of concentration to produce the greatest intensity of flood discharge.

Prior to 1999, the most authoritative study of floods in the UK was the *Flood Studies Report* [5] published in 1975. A *Guide to the Flood Studies Report* was published by the Institute of Hydrology in 1978. The Guide is, however, out of date and no exact replacement exists. Detailed advice on applying the flood studies report to reservoir safety is given in Appendix 2 of *Floods and Reservoir Safety* [8]. The steps in the derivation of reservoir design flood inflow used may be summarized as follows:

(a) deduce appropriate design storm duration for the reservoir catchment, making an estimated allowance for reservoir lag
(b) compute depth of storm rainfall of that duration and of design severity averaged over the catchment
(c) distribute design rainfall in time according to an appropriate profile of intensity
(d) deduct storm losses due to detention and infiltration to derive excess rainfall
(e) multiply the increments of storm rainfall excess, by the unit hydrograph ordinates and cumulatively add the resulting increments, to produce the storm runoff (the flood response hydrograph)
(f) add base inflow (i.e. flow before the onset of the flood) to obtain the complete flood hydrograph.

For the estimation of the probable maximum flood (PMF), the following modifications are advised:

(g) adopting a different profile of storm intensity with time
(h) adding snow-melt (at a uniform rate) to precipitation in winter (up to a mapped total)
(i) reducing the time to peak of the unit hydrograph by one third and so raising its peak flow rate by 50% (based on observations of severe events, instead of using the mean time to peak of a wider range of floods)
(j) storms resulting from probable maximum precipitation (PMP) in winter and summer may be estimated separately because it is not always obvious at the outset which will generate the critical runoff.

Each of these steps was carefully defined in the *Guide to the Flood Studies Report*, but as mentioned, the Guide is out of print.

The concept of PMP is well described in FEH Volume 4 [4], defined by World Meteorological Office (WMO) (1986) as 'theoretically the greatest depth of precipitation for a given duration that is physically possible over a given size of storm area at a particular geographical location (with no allowance made for long-term climatic trend)'. Maps of theoretical maximum precipitations (EMPs) for durations of 2 hours, 24 hours and 25 days enable extreme rainfalls to be estimated for any location and duration in the UK and Ireland [5].

The ICE's *Floods and Reservoir Safety* [8] provides a very helpful and concise guide to its subject, with a summary of the FSR methodology as well as detailed advice on its application and a rapid method of calculation which may be used at existing dams as a first approximation of flood potential. The rapid method has, to some extent, been superseded for those with access to the Institute of Hydrology software MICROFSRV2, which covers the method of the FSR quite fully and includes reservoir routing. Version 2 performs all the calculations necessary for the derivation of the routed design flood hydrograph and the maximum flood surcharge in the reservoir. It has menus to guide users through the recommended technologies (and, where necessary, the alternatives).

Floods and Reservoir Safety [8] deals with the subjects of floods and waves protection standards, derivation of reservoir design flood inflow, reservoir flood routing, wave surcharge and dam freeboard, floods during dam construction and dam improvement works and the overtopping of embankment dams, among other related subjects, including a glossary, within a total of some 70 pages — perhaps the most concise guide on any subject known to the Author.

The Institute of Hydrology publication *Reservoir Flood Estimation: A New Look* [9] provided a useful commentary and includes a survey of the then current procedures, a section on reservoir routing and a selective review of procedures and developments in six other countries. It points out that some of the methods given in the FSR had been subject to revision or extension and it was necessary for users also to refer to the *Flood Studies Supplementary Report* (FSSR) series. To May 1992, there had been 18 FSSRs and of particular relevance to reservoir flood estimation were FSSRs Nos. 10 [10] and 16. FSSR10 [11] sets out a method for calculating flood estimates for reservoirs in cascade. FSSR16 presents revised parameter estimation equations for the FSR rainfall–runoff method of flood estimation, on which flood estimates in the UK were to be based.

The intention of the FEH was largely to supersede the FSSRs by consolidating all the practical advice on the rainfall–runoff method which was issued in the FSR and FSSRs. The FEH was prepared in a way which should obviate the need for a guide in a form similar to that which was found desirable and published for the FSR in 1978 (see FEH Volumes 1, 1.2 and 1.3).

Flood studies: second deterministic method

Envelope curves have been produced for many countries and regions to show the greatest intensities of flood discharge ever recorded or estimated on the basis of historic records, for any river in the country or region concerned, in relation to catchment area.

Examples for catchments in the Philippines and Indonesia are given in Figs 6.1 and 6.2 and for mean annual floods in Visayas, Philippines in Fig. 6.3. It should not be assumed that the envelopes include the maximum possible or maximum probable floods, because the records on which they are based may go back only tens of years, not the hundreds or thousands of years which would be necessary to

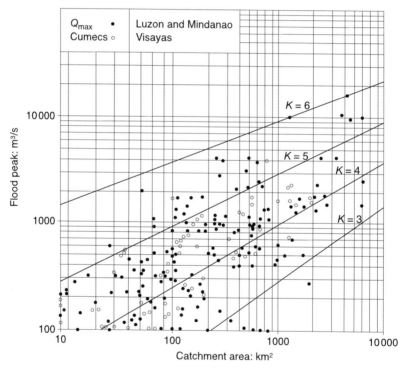

Fig. 6.1. *Maximum recorded flood peaks, Luzon, Mindanao and Visayas, Philippines.* $PMF = (A/10^8)^{1 \cdot 0 - 0 \cdot 1 K}$ *(in $10^5 \, m^3/s$). IANS Publication No. 131, World Catalogue of Large Floods. Note: K is a scaling factor and varies between 2 and 8 (Francou Rodier equation). An envelope curve for the world's largest floods has* $K = 8$

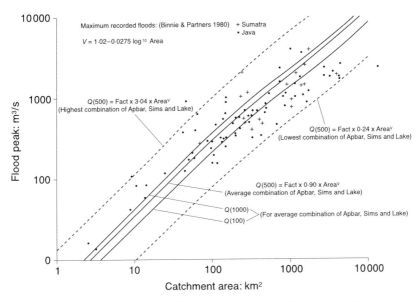

Fig. 6.2. Maximum recorded floods in Java and Sumatra and regression equation estimates

avoid excessive extrapolation. The data plotted on the three figures illustrate the wide variation of large floods liable to be experienced for any given area of catchment, even in the same region, and also illustrate, with some indication of reliability, how envelope curves can serve as a

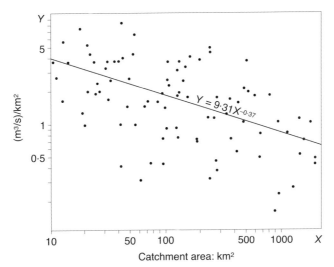

Fig. 6.3. Unit mean annual flood as a function of catchment area, Visayas

guide to assessment of the possible limits within which future flood intensities can be expected to lie, based on historic records. Flood frequency curves for Java and Sumatra have also been published in Farquharson, *et al.* (1987) [12] and Meigh, *et al.* (1997) [13] in a slightly modified form in each case.

Flood studies: first probabilistic method

As already mentioned, there is no hard and fast division between deterministic and probabilistic methods of flood estimation, and most methods contain at least an element of both. However, in either case, methods based on a historical record produce only extreme values of droughts and floods which are characteristic of that record and do not analyse the potential for values of events which are 'outliers' (values outside the characteristics already observed). This applies to probabilistic time series but these do attempt to derive the probabilities of extreme values much beyond the length of record.

A basic understanding of statistics is needed in the application and use of extreme value distributions. There are a number of *time series distributions* which have been developed to predict the probability of floods, not only floods of very low probabilities which can be regarded as extreme values, e.g. probabilities of 1 in 10 000, which on average are not expected to be exceeded more than once in 10 000 years (which have a probability of about 1 in 100 of not being exceeded more than once in a project life of 100 years) but also to predict floods of much greater probabilities, 1 in 100 or 1 in 30 or 1 in 20 or 1 in 10, which may be considered sufficiently unlikely to occur during much shorter critical periods or when and where a higher risk is acceptable, for example during construction, which may be usually 1, 2, 3, or possibly more years. The time series distributions are fitted as well as possible to the recorded values and are used to extrapolate the record as far as required to estimate the probabilities of floods outside the recorded values. Here lies the weakness of the method, because it is tempting and deceptively simple to make long extrapolations whereas it has been demonstrated that extrapolations become increasingly unreliable the longer they are in relation to the length of record. One of the early series to be developed was Gumbel's extreme value distribution. As the name implies, it was systematically developed for the prediction of extreme values, particularly floods. It is still much used and not superseded. However, in some cases other distributions are available which may or may not be found to make a better fit to the historical data, for example the Pearson Normal and the Pearson Log Normal Type 3 distributions. Distributions for flood frequency analysis are outlined in Chapter 15 of Volume 3 of the FEH, including the Generalized Logistic (GL), recommended for the UK and the Generalized Extreme Value (GEV), described because of its theoretical and historical importance. The Gumbel distribution is a particular case of the GEV distribution which was adopted by the FSR. The extreme value distributions just referred to are applicable to the analysis of series of annual maxima flood peaks. Other distributions are suitable for fitting to points over a specified threshold (PoT). Missing data in the series in either case must be infilled to provide a continuous and consistent basis for analysis.

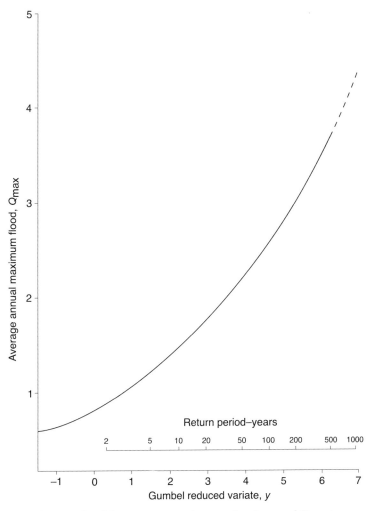

Fig. 6.4. Average flood frequency growth curve for Java and Sumatra

Consistent results of an extreme value analysis using the probabilistic methods outlined above can be expressed as a flood frequency curve specific to the site of interest.

Flood studies: second probabilistic method

This comprises are flood growth curves which express flood discharges for a range of probabilities, as multipliers of the average annual maximum flood (QMAF) or of the median annual maximum (QMED), plotted against probability. Examples of the former from Indonesia and the Philippines are shown in Figs 6.4 and 6.5. The probabilities have to be derived from time series distributions. The QMED was adopted for the purposes of the *Flood Estimation Handbook* for the UK (Volume 3, Chapter 11, 3.4). Values of RMED (the median of annual maximum rainfall at a site) were interpolated between rain gauge sites using topographic information, giving 1 km grids of RMED covering the UK (FEH, Volume 2, Chapters 6, 7 and 8).

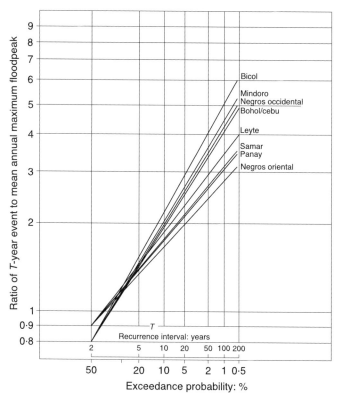

Fig. 6.5. *Regional flood growth curves, Visayas*

These were used to derive rainfall growth curves. However, as described, in October 2000 this particular application was still subject to validation. Generalized flood frequency curves derived from growth factors for application within analagous geographic regions elsewhere should be used only with caution unless satisfactorily validated. However, such curves still find application overseas.

Droughts

It should be possible to proceed from the extensive work done on probable maximum floods to the corresponding case for droughts. However, the problem is compounded by the need for drought minima that cover from a few days up to a decade or more. The probable worst drought (PWD) has been defined as the lowest possible volume of flow over any stated duration that can be realistically

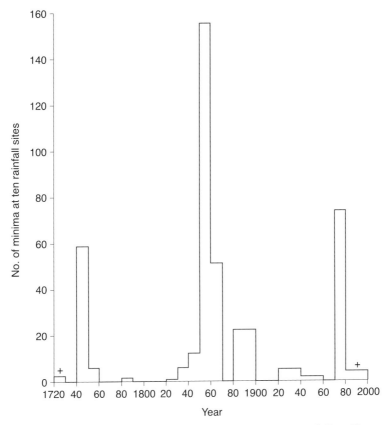

Fig. 6.6. Decades (by end date) in which rare droughts occurred (3 to 60 month rainfall droughts)

expected from an actual combination of meteorological and natural catchment circumstances; it is a lower bound, approached asymptotically by historic recorded flows and is rarer than, say, a 1 in 10 000 chance event.

Britain and Australia led the way in early drought probability analyses from about 1950. However, neither that work nor the Institute of Hydrology (IoH) Low Flow Studies ever went to events rarer than 1 in 100 years. Strides have been made in recovering data remaining from past rare long dry periods, as in the 1850s (Fig. 6.6), spurred on by the outstanding 1988–92 drought in the east of the UK. Figure 6.6 seems to indicate the occurrence of exceptional intensities of drought about every 110 years.

The UK has not published a lower envelope of recorded lowest rainfalls but an outline attempt has been presented by F M Law of the IoH, as shown in Fig. 6.7. F M Law suggested that this, leaving climate change apart, is unlikely to be far from the PWD if it could occur as a single event. The routing of that rainfall minimum, in dimensionless form, through a lumped storage representing any catchment system of concern, should be able to give the ultimate view of the lowest runoff. Research in this field is seriously lacking, as more attention has been

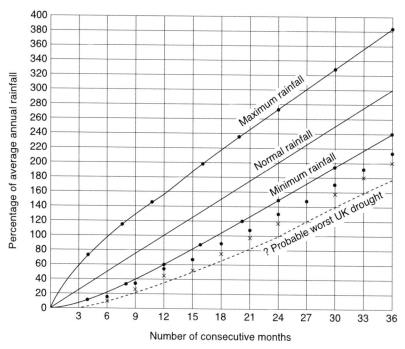

Fig. 6.7. Extremes of rainfall for groups of consecutive months

paid to advanced shortage operation rules than to the event which may test them to breakpoint. Since it is droughts which govern the need at critical times to supplement natural supplies, one may well say that more attention should be given to this subject.

In conclusion, the following extract from the *New Scientist*, 20 August 1994, speaks for itself.

> After predicting a cool, wet summer back in March, Japan's Meteorological Agency has at last revised its long-term forecast. On 20 July, after many rivers had been dry for weeks and in many homes the taps were only turned on for five hours a day, the agency warned that the summer would, after all, be hot and dry.

References
[1] McMahon, T. A. and Mein, R. G. (1978) *Reservoir Capacity and Yield*. Elsevier, Amsterdam.
[2] Morris, G. L. (ed.) and Jiahua Fan (1997). *Reservoir Sedimentation Handbook. Dams, Reservoirs & Watersheds for Sustainable Uses*. McGraw-Hill, New York.
[3] Institution of Engineers, Australia (1999). *Australian Rainfall and Runoff*, eight volumes, 1st Report (1958), The Institution, Sydney.
[4] Reed, D. *et al.* (1999). *Flood Estimation Handbook*. Institute of Hydrology, Wallingford, UK, five volumes, 1st Edition.
[5] Institute of Hydrology and Natural Environment Research Council (NERC) (1975). *Flood Studies Report* (FSR), Institute of Hydrology, Wallingford, UK, five volumes.
[6] Babtie Group in association with the Centre for Ecology and Hydrology (CEH) (2000). Wallingford & Rodney Bridle Ltd *Report to DETR: Reservoir Safety — Floods and Reservoir Safety — Clarification on the use of FEH and FSR design rainfalls*, pp. 1–38. Bothwell Street, Glasgow.
[7] DETR (2000). *Background and Departmental Response to the September 2000 Research Report on Reservoir Safety*, pp. 1–9. DETR, London.
[8] Working Party on Floods & Reservoir Safety (1996). *Floods and Reservoir Safety*. Thomas Telford, London, 3rd Edition, pp. 1–63.
[9] Institute of Hydrology and Natural Environment Research Council (NERC) (1975). *Estimation of floods by unit hydrograph method*, Institute of Hydrology, Wallingford, UK, five volumes.
[10] *Flood Studies Supplementary Report 10: A guide to spillway flood calculations for a cascade of reservoirs*, 1983.
[11] *Flood Studies Supplementary Report 16: The FSR rainfall–runoff model parameter extraction equations, updated 1985*.
[12] Farquharson, F. A. K. *et al.* (1987). Comparison of flood frequency curves for many different regions of the world. *Regional Flood Frequency Analysis*. Singh V. P. (ed.), Reidel, Dordrecht, pp. 223–256.
[13] Meigh, J. R. *et al.* (1997). A worldwide comparison of regional flood estimation methods and climate. *Journal des Sciences Hydrologiques* **42**, No. (2), 225–244.

Addendum to Chapter 6: Guidelines from the United States of America

The United States Army Corps of Engineers Hydrologic Engineering Center has produced software packages in an HEC series 1–6: these are widely used and provide valuable methods of modelling the hydrology and hydrogeology of catchments including lakes and reservoirs in series down a valley, and covering reservoir operation simulation. A summary is given in the following table but it should be noted that, in the past, some problems have been encountered in the use of these programs in the UK. The cost of each program in the UK was US$ 65 to US$ 200, including manuals and diskettes, as listed in the table.

Table 6.1. Methods of modelling the hydrology and hydrogeology of catchments (US Army Corps of Engineers Hydrology Engineering Center)

Reference	Title	Manuals, diskettes & cost in UK	Summary
HEC-1	Flood Hydrograph Package	3 manuals, 5 diskettes, 6 supplements, $160	Flood hydrograph computations associated with a single storm
HEC-2	Water Surface Profiles	3 manuals, 4 diskettes, 2 supplements, $160	Profiles for one-dimensional, steady, gradually varied flow
HEC-3	Reservoir System Analysis for Conservation	1 manual, 1 diskette, $65	Multi-purpose routings for a multi-reservoir system for water supply and hydropower
HEC-4	Not available to the Author	Not known to the Author	Not in list available
HEC-5	Simulation of Flood Control and Conservation Systems	8 manuals, 9 diskettes, 4 supplements, $200	Sequential operation of a reservoir–channel system with branched network configuration
HEC-5Q	Simulation of Flood Control and Conservation Systems	5 manuals, 3 diskettes, 4 supplements, $200	Sequential operation of reservoir systems for flood control and conservation purposes
HEC-6	Scour and Deposition in Rivers and Reservoirs	2 manuals, 4 diskettes, 2 supplements	One-dimensional sediment transport model to calculate water surface and sediment bed surface profiles

7. Spillways

Intensity of flood flow to be provided for

It is a design necessity to specify a flood, in combination with wave action, which a dam must be capable of withstanding. The passage of this flood through the reservoir should cause no fundamental structural damage to the dam which could endanger lives and result in flooding of valuable property. International Commission on Large Dams (ICOLD) Bulletin 61 [1] advises:

- that a dam, with its foundation and environment, should most economically perform satisfactorily its function without appreciable deterioration during the conditions expected normally to occur in the life of the structure; small limited displacements, surface cracking and other minor deterioration can be tolerated
- a dam must not fail catastrophically during the most unlikely but possible conditions which may be imposed. Extensive distortion and cracking and even appreciable permanent movement, deformation, and seepages requiring repair may be permitted, providing that safety is ensured.

The ICE publication *Floods and Reservoir Safety* (see reference [8] in Chapter 6) provides a well-used table which illustrates most of the typical options and some of the quandaries on flood, wind and wave standards (Table 7.1). The contents of the table are described in detail below.

(a) *The first two columns.* Four categories, A to D, in a reducing scale of potential effects of a dam breach, as described. ICOLD Bulletin 72 [2] also adopted four dam categories but proposed a system of classification on a numerical basis depending on the volumetric capacity of the reservoir, the height of dam, the number of persons who would be at risk in the event of failure of the dam, and the potential downstream damage which failure would cause. These factors were adopted in classifying dams in *An Engineering Guide*

Table 7.1. Flood, wind and wave standards by dam category (from Floods and Reservoir Safety, ICE, 1996)

Dam category	Potential effect of a dam breach	Initial reservoir condition	Reservoir design flood inflow		Concurrent wind speed and minimum wave surcharge allowance
			General standard	Minimum standard if overtopping is tolerable	
A	Where a breach could endanger lives in a community	Spilling long-term average inflow	Probable maximum flood (PMF)	10 000-year flood	Mean annual maximum hourly wind speed Wave surcharge allowance not less than 0·6 m
B	Where a breach: (i) could endanger lives not in a community or (ii) could result in extensive damage	Just full (i.e. no spill)	10 000-year flood	1000-year flood	Mean annual maximum hourly wind speed Wave surcharge allowance not less than 0·6 m
C	Where a breach would pose negligible risk to life and cause limited damage	Just full (i.e. no spill)	1000-year flood	150-year flood	Mean annual maximum hourly wind speed Wave surcharge allowance not less than 0·4 m
D	Special cases where no loss of life can be foreseen as a result of a breach and very limited additional flood damage could be caused	Spilling long-term average inflow	150-year flood	Not applicable	Mean annual maximum hourly wind speed Wave surcharge allowance not less than 0·3 m

Notes: 1. Where reservoir control procedures require, and discharge capacities permit operation at or below specified levels defined throughout the year, these specified initial levels may be adopted providing they are stated in the statutory certificates and/or reports for the dam.
2. A method for deriving the mean annual maximum hourly wind speed is given in Section 5.2.
3. See Section 5.6 for the determination of minimum dam freeboard for general standard dams.
4. For reservoirs without direct catchment, a spillway may not be required provided that the dam freeboard is adequate to contain the appropriate wave surcharge recommended in Section 2.6.

to Seismic Risk to Dams in the UK (see reference [3] in Chapter 9). The standards for dam spillways adopted by the US Army Corps of Engineers are also comparable and include reservoir capacity and dam height but different values of the parameters are used. The objective is to arrive at a design which is acceptably safe but not unnecessarily expensive in its provisions for flood discharge capacity. In theory it would be possible to design a dam for which the present value of the cost of the dam added to the present value of the probable cost of damage due to floods during its lifetime, would be a minimum (i.e. optimum value). This approach is not, however, very practicable because it was thought (by the Chairman of the ICOLD Committee on Dam Safety among others) that it would not be acceptable to put a monetary value on human life. This thinking would lead to the adoption of extreme loading in the design of dams for large reservoirs; extreme loading would never be exceeded as far as could be foreseen. Such a specification might be avoided if contingency plans (emergency preparedness plans) were prepared which would virtually guarantee that no lives would be lost in the event of failure. Emergency preparedness plans are being put into effect at many dams.

Publication of a study by Construction Industry Research and Information Association (CIRIA) [3] on Risk and Reservoirs (CIRIA RP568) was awaited at the end of 2000. Furthermore the ICOLD Committee on Dam Safety had started work on a new bulletin *Guidelines on Risk Assessment for Dams* [4]. It was to be approved for publication in 1999. The Author hopes that efforts are being made to achieve compatibility between the two related publications. Consistent and generally accepted standards are badly needed.

(b) *Column three: Initial reservoir condition.* Clearly, the less full the reservoir is at the onset of a flood, the lower the flood level should rise, but if the volume of reservoir storage is insufficient to attenuate the peak flood discharge, the reservoir must rise to such a level that the maximum discharge out of the reservoir is equal to the maximum intensity of flood inflow, when the reservoir level ceases to rise. At this point, what flows in must flow out because storage used is not increasing.

(c) *Column four: Reservoir design flood inflow.* Three standards are suggested. The probable maximum flood (PMF) is, in my view, misnamed because it has no probability attached to it — it is the flood hydrograph resulting from the worst flood-producing catchment characteristics that can be realistically expected in the prevailing meteorological conditions. It results from the (theoretical) greatest depth of precipitation meteorologically possible for a given duration, for a given basin, at a particular time of year (i.e., it results from probable maximum precipitation (PMP) (Chapter 6)). In other words, it is just possible, but should be incapable of being exceeded. Attempts have been and are being made to relate a probability of being exceeded to PMF in any one year, which it is necessary to do for purposes of risk analysis, but to do so implies a probability curve which asymptotes to the horizontal at its upper end, giving a single value of discharge for the lowest probabilities. Two methods are indicated in Volume 4 of the FEH (see reference [4] in Chapter 6): the first is in *Australian Rainfall and Runoff* (see reference [3] in Chapter 6) and the second was developed for the UK by Lowing and Law [5].

In the UK, the then DETR considered that guidance on floods and reservoir safety should move away from an approach based on estimates of PMP towards an exclusively 'T-year' approach, with flood safety becoming absorbed into a more integrated approach to all aspects of reservoir safety. The Author expects that this proposal will be subject to difficulties and differences of opinion within the community involved with dams and reservoirs.

It can be noted, that where the dam poses lower hazards, Table 7.1 proposes less intense flood inflows for design, the least intense being that which has, on average, a return period of 150 years, i.e. has a probability of being exceeded in any one year of approximately 1 in 150. It may be noted that a '1 in 10 000' year flood has a probability of being exceeded in any one year of approximately 1 in 10 000, equivalent to a probability of 1 in 100 of being exceeded in the life of the dam, if the life is taken as, say, 100 years.

For rapid exploratory assessment only, the T-year event in the UK may be computed as of the order of a fraction of the PMF as shown in Table 7.2.

The lower standards indicated 'if overtopping is tolerable' are also subjective. The resistance of dams to overtopping has been

Table 7.2. T-year event in the UK for rapid exploratory assessment

T-year event	Equivalent fraction of PMF for rapid assessment only
PMF	1·0
10 000-year	0·5
1000-year	0·3
150-year	0·2

much under study both here, by the Construction Industry Research and Information Association (CIRIA), and in the United States of America by the US Corps of Engineers, by the American Society of Civil Engineers (Hydraulics Division), by the US Department of Agriculture, and by the US Bureau of Reclamation. For all dams, the appropriate 'general standard' of flood should be routed through the reservoir. In the case of an existing dam, where it is found that this causes overtopping of the dam to occur, the capability of the dam to withstand the overtopping should be assessed from the guidance given in the ICE guide. The overflow caused by a general standard design flood inflow must be limited both to the intensity and duration which will not cause unacceptable damage. Where overtopping is assessed to be tolerable, the dam should be made capable of passing the minimum standard design flood inflow safely *without* overtopping.

A distinction should be made between overtopping by 'still water', and wave overtopping and carry-over of water by wind. The results of work on these aspects are growing.

(d) *Column five in Table 7.1: Concurrent wind speed and minimum wave surcharge allowance.* The values given are also subjective and reasonable for the UK, in accordance with experience. However, joint research has been in progress at Sheffield University with the Institute of Hydrology, to analyse the joint probability of high winds and floods of various probabilities, together with varying initial reservoir conditions. Thus it was hoped that it would now be practicable to give the probability for the combined conditions during floods given in Table 7.1. Clearly the probability of a combination of severe conditions occurring together must be lower than the probability attached to the single flood inflow component. However, a realistic approach to assessment on this basis has not yet been developed.

Finally it should be noted that the standards of safety of the dam and the spillway need not necessarily be the same: for example, overtopping of spillway side walls might be acceptable, providing that the spillway could continue to discharge the flow required and providing that its overtopping could not lead to a serious failure of the dam and providing that the damage could be satisfactorily repaired after the flood receded.

The effect of type of dam on type of spillway needed

The desired type of dam exercises constraints on the type and cost of spillway and vice versa. The final choice of each must provide a technically sound combination of dam and spillway but usually is decided by the optimum economic combination. The main forms of spillway which are available, their characteristics and suitability for different types of dam are discussed below.

Open spillways

These normally allow uncontrolled free flow over the crest of a weir or dam. Because they do not depend on timely and trouble-free operation of gates or valves, properly designed open spillways are perhaps the safest and most reliable means of flood discharge. Once the reservoir level rises above the crest, overflow commences. Since overflow is liable to spill to waste, open spillways are usually designed for relatively low unit discharges (discharge per unit length of crest) so that the flood rise is not excessive and does not store a large volume in the reservoir which cannot be retained, and so must be wasted. Thus open spillways are most suitable for valleys that are wide in relation to the potential flood rise.

The surface of the crest must be erosion resistant and is normally of good-quality concrete or masonry or might be of RCC. The crest must be accurately formed and designed to intrude slightly into the lower nappe of the overflowing jet at maximum discharge, in order not to allow a void between jet and crest with resulting possible sub-atmospheric pressures and cavitation. Design charts for the design of such crests to achieve the maximum unit discharge are readily available. The dispersal of the energy of the overflowing jet is a major design consideration which will be dealt with presently.

Very often bridges are formed across open spillways to provide road access across the dam during floods or for access across the river. They can be supported on piers, which should be designed with rounded noses and streamlined tail shapes, in order not to reduce the overall unit discharge capacity of the spillway, unnecessarily. In any case, estimation of the discharge capacity must allow for end contractions and friction losses at each abutment and at the piers.

Open spillways can be readily and economically formed to provide the crests of masonry and concrete dams. They can be formed from RCC dams but are probably best surfaced with high-quality, conventionally vibration-compacted concrete, using formwork (shuttering), 'screeding' and 'floating' to create the designed profile. They can be incorporated in concrete or masonry buttress dams or hollow gravity dams but the surface of the crest needs to span between buttresses, by the use of arches or reinforced concrete beams or slabs, in order to prevent the jet falling inside the dam at low discharges where it could cause erosion and cavitation. There are cases where open spillways have been incorporated in embankment dams by paving the crest and downstream surface of fill with, e.g., interlocking paving slabs or beams [6]. Care has to be taken to ensure that the paving is sufficiently flexible and deformable to follow the deformations and settlement of the supporting fill, without allowing leakage which could erode voids in the embankment (that is, cause internal erosion). The joints in the slabs or beams have to be sealed against such leakage. Thus, overtopping of embankment dams involves quite meticulous design and is usually not adopted for high unit discharges. The integration of mass concrete, masonry, or RCC spillways in embankment dams is also to be avoided where practicable, because of the difficulty of designing the interface between fill and spillway, to avoid leakage which might lead to internal erosion in the fill. To avoid cracks in the embankment or along the interface due to tensile strains, the design should be such that the water pressure from the reservoir compresses a sealing layer in the fill against a supporting profile in the flank of the spillway. Wherever seepage or leakage into fill is foreseeable (through beams or paving or past an interface) sand or gravel or geomembrane filters should be designed to intercept fine material leached from the fill by the leakage; drainage of the escaping water should be provided for.

Bellmouth spillways are sometimes used to discharge flood water over a continuous, usually circular, crest and down a shaft into a conduit. They are sometimes known as 'morning glory' spillways. The points to note are that the entrainment of air, which could have explosive effects, is to be avoided by careful design and, if necessary, by model experiments, and that the maximum discharge capacity will have a physical limit, likely to be governed not by the reservoir level but by the hydraulic conditions in the shaft.

Gated crest spillways

Gated crest spillways can be provided in place of open spillways but are most economical and suitable where it is desirable or necessary to minimize the height of rise in reservoir level during floods (or even to prevent any rise) and to store as much as possible of the inflow to the reservoir during floods. The high cost of hydraulic gates means that the area of waterway closed by gates has to be kept down to an economic minimum. The height to width ratio of gates is usually in the range of about 1:1 to 1:2. Water can normally be stored in the reservoir up to the top level of the gates. With further increase in flood inflow, one or more of the gates can be gradually raised (or lowered) to allow discharge, to prevent the rise in water level from exceeding the designed limits. Alternatively, in some cases gates are designed to allow overflow, or the crest section of the gate can be lowered by rotation or sliding, to allow overflow.

There are numerous alternative types of gates (described below) from which a technically sound and economic choice can be made, to suit the characteristics of the site and operational requirements. In all cases they must be designed to avoid vibration (which can lead to failures due to resonance) during operation. This requires suitable rigidity and hydraulic profiling of overflow and/or underflow surfaces.

(a) *Vertical lift gates*, suspended from piers or bridges or gantries on the bridge and operated by hoists. These can usually be lifted clear of the discharging jet of water and consequently necessitate high piers (to allow for the hoisting tackle and gate height). They should be designed for operation without vibration also at partial openings, in case the gate operation is intentionally stopped, or stopped due to defect.

(b) *Radial gates*, sometimes called 'Tainter' gates, supported by radial arms hinged on either side at the piers or abutments. Operation can be by hoists in or on the piers or abutments. This is perhaps the most economic type of gate for medium or large spans and is suitable for spans up to 15 or even 20 m. The gates can be designed for raising clear of the discharging jet of water or for lowering into a recess in the crest of the dam. They should, like vertical lift gates, be designed for operation at partial openings.

(c) *Drum gates*, supported by horizontal hinges on the crest. These gates are an overflow type, and as the name implies, formed by a totally enclosed structural membrane: they are raised or lowered by flotation in water in a recess in the crest of the weir. Water is introduced into the recess to raise the gates or drained out of the recess to lower them.

(d) *Bear trap or flap gates*, supported by horizontal hinges on the crest. They are raised from or lowered into a recess in the crest of the weir by hoist or hydraulically operated piston. As in the case of drum gates, when lowered, they conform to the profile of the crest of the weir, in order to maximize unit discharge.

The operation of the above types of gates can be achieved manually without power or by electromechanical operation of hoists. Operation can also be achieved automatically, under control by computer, on the basis of reservoir level and gate openings (which together determine discharge), or operation can be achieved automatically under hydraulic control using the same criteria. In any case, since operation of heavy gates by hand, without power, is extremely slow, an alternative source of power is very desirable as stand-by in case of emergency. For example, a diesel generator might provide suitable stand-by power in case mains power should fail during a flood — a quite likely occurrence. For safety during floods, it is also desirable to have spare gate capacity, in fact one or more spare gates, in case one or more gates should be out of operation during floods. It may be noted that it follows that the total number of gates should be not less than two in parallel.

It is necessary to provide for the installation of stop logs, or bulkhead gate(s), upstream of hydraulic gates, so that the gates can be isolated from the reservoir and maintained or repaired 'in the dry', without having to lower the reservoir level (which may not be practicable).

(e) *Hydroplus gates*, a novel gate design developed by eminent French designers. The gate operation is fully automatic and is controlled without any power by its geometrical shape: it is in the form of a drum which fills with water during floods while at the same time, it is discharging water. The last stage is on reaching a critical high reservoir level when the gate overturns and washes away downstream. The concept is that the gates are inexpensive and it is not envisaged that they would be salvaged and repaired. It is considered that the design and installation should be such that all the gates do not overturn and wash away simultaneously, as that could cause a dangerous flood downstream. Such staging of operation is not difficult. Snell *et al.* [7] have provided a good description of this type of gate.

Low-level flood discharge outlets

Obviously, these outlets, below water level, have to be controlled by gates or valves. They are discouraged for flood control by some authorities for reasons of safety on the grounds that access below water level for maintenance is difficult and the gates and/or conduits may become blocked or seized by sediment deposits or sunken debris. In the case of a new dam, it does, however, seem economically desirable to make permanent use of diversion tunnels which were used for flood discharge during construction. They have a potentially high discharge capacity under full reservoir head. The Portuguese authorities, in particular, have sensible conditions under which low-level gated spillways could safely be used: they require that arrangements should be provided for emptying the reservoir without harm to the dam; bottom- and medium-level spillway outlets would have to be equipped with at least two gates, across separate spans, and it must be possible to operate the gates both locally and by remote control, by means of power having two distinct sources, apart from it being possible for them to be operated manually. There would need to be two gates in each span: one of the gates is to function as a safety gate and the other for normal operational service. The safety gates can be bulkhead or other type and must be able to isolate the gates from the reservoir for maintenance or repair or there must be an overhead chamber into which the gates can be withdrawn, for maintenance or repair 'in the dry'. If the gates or valves are located at the downstream end of the conduit, the isolating gate should be at the upstream end in order to be able to empty the conduit for maintenance. This condition is particularly important in embankment dams where leakage from the conduit into the embankment could cause internal erosion and possible failure of the dam. In embankment dams, the conduit should normally be kept empty, otherwise it would be subject to high pressure from the reservoir and consequent potentially high leakage into the fill in case of rupture. The gates or valves should be operated at least once every year to test their serviceability.

It follows from what has just been said that medium- or low-level outlets should be provided with coarse screens at their upstream end, to avoid the risk of blockage by submerged refuse. Electromechanically operated rakes should be provided for cleaning the screens, the operating controls being readily accessible and facilities being provided for removing from the dam the rubbish removed by the rakes.

The types of gates or valves used to control low-level spillway outlets must be suitable for use under the maximum head from the reservoir — although the probable head when repairs and maintenance are undertaken may be somewhat less. Vertical lift wedge type gates, roller-mounted gates and radial gates, Howell Bunger and 'hollow jet' valves are suitable for operation under high head and, if suitably designed, for operation at partial openings. Conduits are usually designed so that gates or valves downstream can be closed, so that the conduits upstream of them can be filled under full hydrostatic head. Under this condition, the guard gates or valves upstream can be of a less expensive (e.g. roller or sliding) type that can be operated only under balanced pressure, i.e. upstream and downstream pressures being the same. Reference will be made to energy dispersal presently. Energy dispersal can be a serious and possibly unsolved explosive problem, if jets under high pressure are discharged into submerged conduits.

A combination of types of spillway may be used. However, to minimize the cost of the permanent spillway, a temporary emergency or 'fuse-plug' spillway is sometimes used to discharge the peak of an extreme flood, in excess of the discharge capacity for which the permanent spillway is designed (see also under remedial works, p. 300).

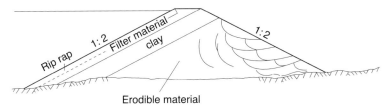

Fig. 7.1. Sketch showing principle of auxiliary fuse-plug spillway

In concept, the fuse-plug spillway is a temporary dam or cut-off in the reservoir rim which would wash away when the flood raises the reservoir level to a predetermined threshold. A conduit, or erosion protection, is required to carry the flood discharged, without unacceptable erosion. The design and intended discharge of the fuse-plug should be decided by the minimum present value of permanent and fuse-plug spillways combined, including the cost of rebuilding the fuse-plug after it is washed away, perhaps more than once (since rebuilding would be required to retain the designed reservoir level). A simple and reliable design of fuse-plug spillway would comprise an embankment with an upstream impermeable clay shell and erosion and wave-resistant upstream face and crest, as shown in Fig. 7.1. The shell downstream would be built of erodible sand or gravel, which would erode and wash away and allow the fuse-plug to collapse, once it were overtopped.

The length of flow path downstream of the spillway crest, to carry the flood discharge back to the river, depends on the siting of the spillway relative to the dam. As already noted, spillways of high unit discharge are best separated from embankment dams. They are usually, therefore, sited on an abutment or separate from the dam, in a saddle on the rim of the reservoir. They therefore usually require a fairly long, lined conduit, with sufficiently high walls to contain the flood to carry it back to the river, or perhaps to a tributary with its confluence downstream of the dam.

Energy dispersal

Energy dispersal of the water overflowing spillways has been the subject of extensive study by hydraulic model experiments, supported by hydraulic analyses. Many authorities, notably the United States Bureau of Reclamation, have drawn up general guidelines (mainly on an empirical basis) derived from hydraulic model experiments [8–10]. Due to the unique geometry and particular conditions at different individual sites, it has been quite usual for specific model studies to be carried out for the design of every new spillway of major dimensions. These studies have been required to improve the hydraulic flow conditions, to improve the energy dispersal efficiency and hence to minimize the cost of the structure. The problems are fundamentally those of supercritical flow, turbulence and wave occurrence. The models have been used to check hydraulic pressures and possible conditions for cavitation to occur, the necessary height of retaining walls, and the discharge capacity of the spillway proposed, among other parameters. In very many cases, the results of such model studies have been published. There is, therefore, a great deal of published information on the subject of hydraulic analysis of the performance of spillways, from which guidance can be obtained on new designs. The need for specific model studies in every case which arises is therefore decreasing although it is still considered good practice to undertake hydraulic model studies for each new major project (since the resulting improvements to the design serve as an insurance against unforeseen problems and may result in savings much greater than the cost of the studies — although this latter cost is not likely to be insignificant).

The following paragraphs summarize the characteristics of the main forms of energy dispersing structures in use.

(a) *A stepped surface to the downstream face of the weir (Fig. 7.2).* This can be created by beams spanning between buttresses to create the downstream face of a dam or can be purpose-made. It has proved successful in causing aeration of the overflowing jet and reducing its energy due to turbulence. Work has been published by CIRIA and others [11,12] to assist in optimizing the dimensions of steps for any design discharge.

(b) *Ski buckets — sometimes called 'flip buckets'.* Buckets can be built on the downstream face or at the end of a spillway to turn the jet and redirect it to discharge beyond the face of the dam, usually into some form of plunge pool, where its energy is destroyed by turbulence. From the reservoir onwards, the jet travels only under the

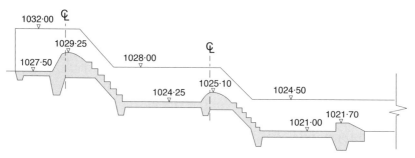

Fig. 7.2. *Stepped spillway (scale 1:400)*

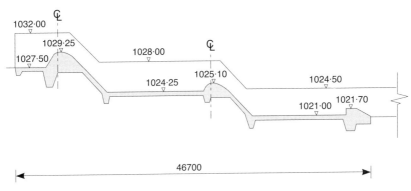

Fig. 7.3. Smooth spillway (scale 1:400)

influence of gravity and friction and wind resistance. For maximum distance of travel through the air, the bucket is usually shaped as the segment of a circle and designed to redirect the jet upwards at an angle of about 45°. The falling jet will eventually erode a deep and extensive hole even in massive rock, so, unless the impact is sufficiently far downstream that the erosion cannot work back (even over a long period of time) to undermine the dam, the plunge pool requires a substantial lining of reinforced concrete. Published work can be used to dimension the bucket and the plunge pool.

(c) *Stilling pools/stilling basins.* Conventionally, the jet falling from an overflow weir is turned to a horizontal direction by a bucket or by the angle of intersection of an apron with the downstream face of the weir (Fig. 7.3). To prevent backward erosion of the dam's foundations, energy dissipation should be constrained to take place on the apron, so that from there on, the flow continues along its natural course at subcritical velocity. This can be achieved by constructing a 'small' weir across the downstream end of the apron which receives the impact of the jet and increases the depth of water, which causes a hydraulic jump to occur on the apron [13]. The length of apron and depth of water on it are critical with regard to the manner of energy dissipation and its performance. Depending on the height of the weir and flow conditions, the end weir can alternatively cause a roller, which travels backwards up the stilling basin and disperses the energy in a roller action. Possible improvements in the efficiency of energy dissipation can be achieved by indentations in the downstream weir (or sill), i.e. providing regular recesses or gaps throughout its length, and/or by constructing line(s) of baffle blocks across the surface

114 | *Reservoir engineering*

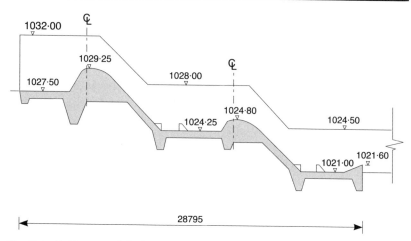

Fig. 7.4. Spillway with baffles (scale 1:400)

of the apron (Fig. 7.4), all of which create irregularity and turbulence in the flow. The end weir and blocks need to be designed to resist both the force and erosion from the impacting jets of water. Figure 7.5 shows a combination of the features described. In this case it was found that the provision of stepped faces to the weirs did not add significantly to the energy dispersal achieved by the sills and baffle blocks.
(d) *Other forms of stilling basin*. Other forms have been developed for particular purposes, e.g. 'impact' stilling basins to 'destroy' the energy of the jet discharging from a pipe or valve under high head, by impact on a transverse wall.

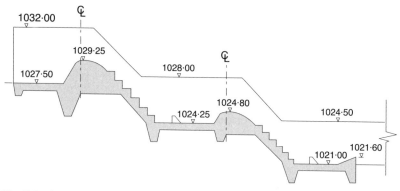

Fig. 7.5. Stepped spillway with baffles (scale 1:400)

Certain types of valve are designed to dissipate the energy of the discharging jet by aeration or turbulence, dispersal or concentration. Examples are needle valves, which concentrate the jet and are expensive (for heads up to 200 m high and up to 2 m diameter) and Howell Bunger valves (see Fig. 12.7), which disperse the jet widely and are economic and effective. It is important to ensure that there is facility for a sufficient flow of air (often considerable) to the point of discharge from the valve. Butterfly valves (perhaps the most economic form of high head valve), hollow-jet valves and others serve many purposes but not that of energy dispersal. The subject of gates and valves [14, 15] is returned to in Chapter 12.

The surfaces over which jets at supercritical velocities flow, should be smooth and erosion resistant, otherwise cavitation may be initiated by irregularities. At velocities liable to cause cavitation, this can be avoided effectively and economically by introducing air-entraining ducts, steps and/or sills in the surface over which the jet travels. Guidelines for the geometry and dimensions of such provisions have been published. It is likely to be necessary to provide for a sufficient flow of air to the ducts, steps or sills, by conduits for the air, which can be built into, e.g., the vertical sides of the spillway, with provision at the entrance for cleaning the ducts and prevention of debris entering.

References

[1] International Commission on Large Dams (eds). *Dam design criteria, the philosophy of their selection.* Bulletin 61, ICOLD, Paris, 1988.

[2] International Commission on Large Dams (eds). *Selecting seismic parameters for Large Dams.* Bulletin 72, ICOLD, Paris, 1989.

[3] Construction Industry Research and Information Association. *Risk Management for UK Reservoirs* (CIRIA RP 542, 2000).

[4] International Commission on Large Dams (eds). Committee on dam safety: guidelines for risk assessment for dams (in preparation).

[5] Lowing, M. J. and Law, F. M. (1995) *Reconciling flood frequency curves with the Probable Maximum Flood.* Proc. British Hydrological Society, 5th National Hydrology Symposium, Edinburgh, vol. 1, pp. 3.37–3.44.

[6] Berry, N. S. M. *et al.* (1994). *Kontagora Dam: aspects of design and construction of an embankment dam relevant to a remote rural area of West Africa.* Proc. Institution of Civil Engineers, Part 1, 1988, vol. 84, pp. 855–874.

[7] Snell, E. F. A. *et al.* (1994). Shongweni to have record ('Hydroplus') fusegates. *International Water Power & Dam Construction*, **14**, 32–38.

[8] US Bureau of Reclamation (eds.) (1987). *Design of Small Dams.* US Government Printing Office, Denver, Colorado, 1987, 3rd edn.

[9] Bradley, J. N. and Peterka, A. J. The Hydraulic Design of Stilling Basins. 6 Parts. *Journal of the Hydraulics Division*, Proc. American Society of Civil Engineers, 1957, October, vol. 83, no. HY5.

[10] Corps of Engineers (eds). *Spillway operation. Flood control by reservoirs.* US Army Hydrologic Engineering Centre, 1976, **7**, ch. 6, February, 14 pages.

[11] Diez-Cascon, J. et al. *Studies on the hydraulic behaviour of stepped spillways.* International Journal of Water Power and Dam Construction, 1991, September, 22–26.

[12] Stephenson, D. *Energy dissipation down stepped spillway.* International Journal of Water Power and Dam Construction, 43, 1991, September, 27–30.

[13] Foster, J. W. and Skrinde, R. A. (1949). *Control of hydraulic jumps by sills.* Proc. American Society of Civil Engineers, **2415**, 973–1022.

[14] Lewin, J. *Hydraulic gates and valves in free surface flow and submerged outlets.* Thomas Telford Ltd., London, 1995.

[15] Gulliver, J. S. and Arndt, R. E. A. (eds). *Hydraulic conveyance design. Hydropower Engineering Handbook.* McGraw-Hill, Inc., USA, 1991, 1st Edition, ch 5, 5.35–5.41.

Further reading

Gordon, J. L. *Vortices at intakes.* International Journal of Water Power and Dam Construction, 1970, April, 22, 137–138.

Mason, P. J. *Erosion of plunge pools downstream of dams due to the action of free trajectory jets.* Proc. Institution of Civil Engineers, Part 1, 1984, May, vol. 76, pp. 523–537.

Sharma, H. R. *Air entrainment in high head gated conduits.* Proc. American Society of Civil Engineers, Journal of the Hydraulics Division, 1976, November, vol. 102, No. 11, pp. 1629–1646.

Water & Power Resources Service (eds.). *Air–water flow in hydraulic structures.* US Department of the Interior, Engineering Monograph 41, 1980, December.

8. River diversions during construction

Intensity of flood discharge to be provided for

The considerations are analogous to those described in Chapter 7. The objective is to provide diversion works which will allow the necessary work to be carried out during construction, without any risk to lives due to the passage of floods, and to keep the cost of the diversion works, and any damage which may be caused by floods, to a minimum. Floods in excess of an intensity for which the works are designed could cause cofferdams to be overtopped and to fail. Early on in the programme for construction, the resulting inundation would submerge excavations in the river bed and its banks, construction plant might be washed away, and cut-off grouting could be interrupted. At a later stage, when concrete in a dam has been placed up to river bed level, floods might pass over the foundations causing only superficial damage, whereas fill placed at low levels could be washed out. In either case, there could be delays to grouting and loss of plant and materials. At a later stage, when the dam has been raised to some height, storage behind the dam might be sufficient to retain the flood, but if, on the other hand, the dam were washed out, remedial works could be very expensive and cause much delay, and there could be risks to lives downstream. Thus the hazards posed by floods during the relatively short period of construction vary with time. A higher standard of safety should be applied during the period when the hazards are the more serious. In the UK, cofferdams which are designed to store a volume of water in excess of $25\,000\,\text{m}^3$ should conform with the requirements of the 1975 Reservoirs Act.

Normally it is accepted that a higher risk can be run during the limited period of construction, than during the permanent life of the dam, when the results of failure could be much more serious. The percentage probability P_r that an event of recurrence interval T years or greater will occur at least once within the next r years can be taken as

$$P_r/100 = 1 - (1 - 1/T)^r$$

For approximate purposes, it may be sufficient to assume that, for a 1 in 10 risk of occurrence during a three-year critical period, the diversion works should be designed to pass a flood with a recurrence interval of not less than 30 years, i.e. the flood return period = duration of risk/risk. If it were thought to be necessary and economic to reduce the risk to 1 in 100 (1%), the return period would need to be increased to 300 years, or if the critical period were only one year, the return period would be 100 years. It is fair to remark that floods more severe than foreseen by estimates of probability have a nasty habit of occurring during the period of construction of dams.

Clearly, contractors have a major interest in avoidance of delays and in safety during construction. However, in tendering, they will be under pressure to keep the cost of such temporary works to a minimum, which could prejudice provisions for safety. The engineer has, however, had more time than the contractor to study the flood risks at the site and should have had all the available hydrological data to hand. In a contract for construction there should be provision for insurances through which the direct cost of damage by flood can be recovered. The cost of consequential losses may, however, be very high and cannot usually be recovered either by the contractor or the employer. Although it is sometimes done, it does not seem sufficient or equitable to leave the whole onus of designing, building and costing temporary diversion arrangements to the contractor. Indeed, it would seem good practice for the engineer to specify in the tender documents that the diversion arrangements shall not be of lower capacity than a certain flood discharge, while contractors (tenderers) are left free to provide for a larger discharge if they so wish. Similarly, it would be good practice for the engineer to include in the tender documents, designs of the diversion arrangements consistent with the design of the dam and to invite contractors (tenderers) to propose alternative arrangements to suit their programme and methods of construction if they so wish. In any case, the contractor should be required to assume responsibility for the design and construction of the temporary works and for the consequences should they prove insufficient, since these would largely affect the contractor's operations and facilities.

Methods of river diversion

In the case of concrete dams, river diversion is often provided for, in the first stage, by dividing the river into two, by cofferdamming one side of the river within which excavation can proceed while the river flows through the remaining open side of the river. A passage for river discharge in the next stage is provided in the dam as it is built in the cofferdammed section of river. This could be a permanent outlet through an embankment dam or simply a gap left in a concrete dam. In either case, provisions must be made for closing this passage, usually by stop logs, bulkhead gate or, in the case of a culvert, by permanent closure valve(s), after river diversion is no longer required. When the section of dam within the cofferdam has been brought up to sufficient height in accordance with the diversion plan, the cofferdam round this section can be removed and the river allowed to discharge through the new gap or culvert. The remaining length of foundations can then be enclosed by cofferdam and excavated, and the foundations of the dam can be built.

In the case of a concrete dam or cut-off trench, it is good practice to carry out the consolidation grouting of the foundations, and the grouting of the cut-off curtain within the foundations, once a modest depth of concrete has been placed. This resists a tendency for the grout pressure to lift the surface of the rock and helps to seal the rock surface and concrete–rock interface. Holes for the grout can be pre-formed in the concrete to save drilling.

An alternative method of river diversion is to construct a tunnel and/or channels for diversion of the river in one bank, in order to carry the river completely past the construction works. It may be possible to drive the tunnel from one portal, to which access is possible in the dry. Usually it is necessary to provide a temporary cofferdam across the upstream portal, so that the intake and closure works can be constructed to the extent necessary, before the river is diverted through the tunnel. It may be noted that diversion by tunnel is particularly suitable in narrow rivers or gorges, where space for dividing the works in the river bed is limited and if the rock in the river bank is suitably sound for tunnelling. After completion of the diversion tunnel and/or channel in the bank, a cofferdam can be constructed across the river downstream of their intake, to divert the water and to close off the river completely, in order to allow excavation and construction in the river bed to proceed unhampered.

The storage capacity created upstream of a cofferdam is usually not sufficient to have a significant effect in attenuating the maximum flood for which the diversion works are designed. The overall head (h) created by the cofferdam on a diversion tunnel or culvert, however, has a significant effect on the potential discharge capacity of the conduit, since the velocity, and hence the discharge, is proportional to $h^{1/2}$. The cofferdam has to be designed as economically as possible but with the same technical and safety considerations as for a permanent dam. The design has to provide against leakage under the cofferdam and through the cofferdam itself. It may be designed to resist erosion by overtopping, to an extent governed by the risk of a flood occurring in excess of that for which the diversion works are designed.

As the construction of the permanent dam proceeds, there will come a stage when the dam itself will serve as a cofferdam, and indeed would do so, if the upstream cofferdam were overtopped. Generally speaking, a dam will be stable during construction if flood water fills against its upstream face without spilling, because the thickest sections of a dam are usually at its lower levels and the water load will be less than that for which the finished dam has been designed. However, chiefly in the case of embankment dams, the risks of the incomplete dam being overtopped and eroded and the risks due to leakage through the dam have to be studied and, if necessary, provided against. In the case of a rockfill dam designed to have an upstream membrane which has not yet been constructed, it may be necessary to provide a light preliminary membrane, perhaps of sprayed mortar (shotcrete). During construction, heavy rocks can be pushed to the downstream face to resist erosion from overtopping, with progressively finer material upstream to prevent small material within the dam from being leached out. Some rockfill dams have been designed to pass flood water safely through the rockfill of the dam itself, as part of the permanent spill arrangements. If the dam is of finer material, it may be necessary to resist erosion in case it is overtopped, by constructing a mattress on the downstream face. This might be of rockfilled 'gabions'. These can be anchored into the fill by steel anchor bars, or the fill may be retained with steel mesh on the face held by steel anchor bars, using 'reinforced earth' techniques.

Closure of river diversions on completion

When the dam has reached a sufficient stage of completion to allow impounding to commence with safety, the Engineer should issue a written certificate to this effect and the procedures required by any statutory regulations (such as the Reservoirs Act, 1975, in the UK) should be implemented. The diversion tunnel/channel(s) or waterway for diversion through the dam can then be closed. This can be effected by permanent gates or valves, if they have been installed and are available for this purpose. Otherwise closure is usually effected at the upstream end by lowering stop logs or bulkhead gate(s), provided for the purpose. Once this is done, the water level upstream is likely to rise quite quickly, because storage in the reservoir at such a low level in the river banks will be relatively small. The closure operation is therefore planned, if possible, for a time of year when low river flows are expected. It may be possible to leave the upstream cofferdam in position, to be permanently submerged by the reservoir, unless it would obstruct and reduce the flow required to an intake to a low level outlet, in which case the cofferdam would have to be lowered or even completely removed. The stop logs or bulkhead gate(s) have to be designed for the differential head which might be applied to them by the level to which water might rise during closure (which rise, therefore, should be as low as can be ensured) and to resist the maximum head to which the reservoir might rise, before a permanent closure can be effected on the downstream side of the stop logs. This closure is often achieved by construction of a concrete plug, constructed by filling a section of the diversion conduit at its upstream end with concrete. Due to the high frictional force, in order, if required, to raise a bulkhead gate under high unbalanced head, it is usually necessary to balance the pressure on both sides of the gate, by admitting water under reservoir pressure to the downstream side, between the gate and the permanent closure. This will probably be effected in time by leakage, but is most reliably ensured by installing a small bypass valve in the gate, with a second guard valve upstream of it. It may sound unreasonably fussy to install two valves for this short-lived purpose, but if the bypass valve fails to open due to some defect, the ensuing difficulty and delay if there is no back-up valve, can far outweigh the extra cost. The two small valves need to be remotely operated. An access manhole through the closure plug is necessary to give access to the bypass valves in case they should need attention. To avoid the trouble and expense of installing two valves, the bypass is in fact often effected by a small slide gate which is built into the bulkhead gate and raised by the first motion of the ropes attached for lifting the bulkhead.

Stop logs are often installed or removed under water, by using a special lifting beam, which automatically becomes detached from the stop log which it is lowering, when the log comes to rest and the load comes off the lifting rope. Similarly, the beam is devised to attach itself to a stop log on to which it is lowered, in order to lift the stop log.

Stop logs and/or bulkhead gates are quite expensive items and it is usually worth going to some trouble to remove and salvage stop logs and/or bulkhead gates, even though they are not required again on the project, and even though it is not essential to do so. It is quite likely that good use could be made of them on some other project if they were to be available. However, the cost of doing so has to be taken into account. In some cases, the slots in which they operate may become deeply submerged in the reservoir and it would not be worth building a high tower on which to mount the hoist for their removal. In such cases they might still be removed by hoisting from a pontoon which could be used to tow the gate or logs into shallow water or on to which they could be lifted, to be floated to the bank.

9. Seismic loading

The relevance of seismic risk to dams in the UK

In view of the remarkably good performance of dams worldwide even under very severe earthquake loading and of the low level, on the international scale, of seismicity experienced in recent times in the UK, it is understandable that most of the population in the UK tend to disregard the risk of earthquakes unless they themselves are responsible for the safety of structures and need to quantify the risks. The magnitudes of the greatest seismic events in the European region since the year 1700 have been $5.8M_L$ in Belgium (Table 9.1, M_L being 'local' or Richter magnitude scale), $5.6M_L$ (or more) offshore and $5.3M_L$ in England and Wales. Figures 9.1 and 9.2 show the distribution of earthquakes in Britain in the last 700 years and the distribution of focal depth of events in the UK and in northwest Europe, which are relatively shallow and hence relatively severe locally at ground level. Magnitudes and focal depths have been greater in northwest Europe than in UK but the trends of magnitude increasing with focal depth have been similar, suggesting that magnitudes as great

Table 9.1. *Ten largest seismic events in provisional catalogue of UK earthquakes 1700–1990*

Date	Magnitude M_L	Location
1938	5·8	Belgium
1931	5·6	North Sea
1977	5·6	North Sea
1926	5·5	Channel Islands
1927	5·4	Channel Islands
1959	5·4	Brittany
1984	5·4	Lleyn
1727	5·3	Swansea
1852	5·3	Caernarvon
1957	5·3	Derby

Source: British Geological Survey Report No. WL/90/28, June 1990

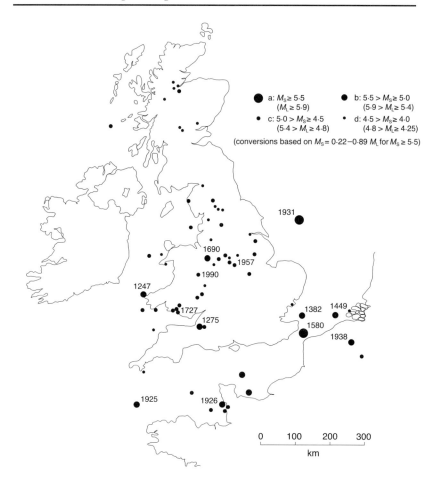

Note: Uncertainty in the epicentral locations is not indicated but is very large in some cases

Fig. 9.1. *Major earthquakes in Britain during last 700 years. (After Ambraseys and Jackson)*

as those in northwest Europe may be realized in the UK, at corresponding depths. The generation of earthquakes is, however, a concept quite different from others with which engineers are more familiar: geologically, it seems that 10 000 years or more ago earthquakes of very large magnitude did occur in this region. The seismic cycle (if there is such a thing) for the UK may be of the order of a million years, so it is unlikely that the maximum value would be found in the relatively short period of historic records. Recent studies have deduced that the maximum magnitude of events in the UK is likely

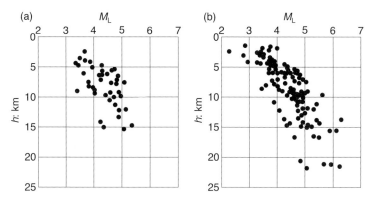

Fig. 9.2. Distribution of macroseismic focal depth with magnitude of earthquake M_L for (a) the UK and (b) northwest Europe [3]

to be more than $5 \cdot 5 M_L$ but less than $7 M_L$. Earthquakes are known to have occurred in other parts of the world not previously regarded as of significant earthquake risk, notably at Tennant Creek in Australia, of up to $6 \cdot 7 M_L$ on 22 January 1988 and at Newcastle, Australia in December 1989 of up to $5 \cdot 7 M_L$. At Newcastle, the total damage, including consequential damage, has been valued at £700 million. The 1884 Colchester earthquake, the most damaging British earthquake in the last 400 years, occurred in an area with no history of previous or subsequent local seismicity. It can be shown (using an attenuation formula by Professor Ambraseys) that an earthquake of magnitude $5 \cdot 3 M_L$ could cause peak ground accelerations of up to $0 \cdot 375g$ or more in the UK, depending on focal distance and ground conditions. This is consistent with independent studies carried out by Long [1] and Irving [2]. Irving's results were adopted in Charles, et al. [4] and give peak ground accelerations of up to $0 \cdot 25g$ in zones of higher seismicity with an annual probability of exceedance of 1 in 10 000 (Fig. 9.3). Ove Arup & Partners [3] assessed the seismic hazard at 12 specific regional locations and Fig. 9.4 shows 11 of these. The range of accelerations at these locations estimated by Ove Arup for an annual probability of exceedance of 1 in 10 000 is up to $0 \cdot 32g$, which is rather higher for that probability than the corresponding figures postulated in [4], which will be referred to further.

The possible loadings in terms of peak ground acceleration which could be imposed on dams by earthquakes as indicated above, are higher than those, usually of less than $0 \cdot 1g$, which have been allowed for in the past by designers of existing dams in the UK. It is therefore necessary for the engineers to satisfy themselves that both old and new dams for which they are responsible could withstand the loading

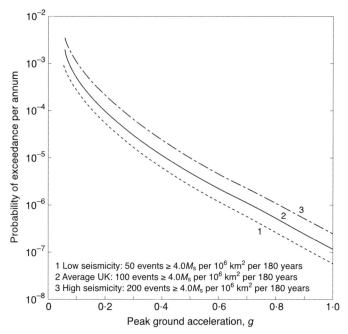

Fig. 9.3. Effect on hazard of variation in the average historical seismicity rate [2]

combinations adopted for design, including the effects of the maximum design earthquake. In doing so, it must be understood that the ground accelerations given are temporarily recurrent, very short-lived peak values over a period of a few seconds, while the dynamic strength rather than the static strength of the structure is the relevant criterion.

Figure 9.5 shows an artificial time history developed by Principia Mechanica Ltd [5] to represent the anticipated nature of response of hard ground to an earthquake in the UK. The response is scaled to a unitized maximum peak ground acceleration of $1.0g$ and it will be noted that the anticipated peaks are attenuated after about 4 s. This subject will be returned to later in relation to applicable factors of safety and how they might be interpreted in the case of seismic loading. More recent practice in the UK has reflected advances in data and understanding: the 1991 guide [4] reported that a peak ground acceleration of $0.3g$ horizontal and $0.2g$ vertical had been proposed for the upgrading in concrete of the 45 m high Upper Glendevon Dam, in an area of medium seismicity southwest of Perth, which was prone to a swarm of events since 1978 of up to $3.5M_L$. For Maentwrog concrete gravity dam in northwest Wales, in an area attributed with high-level seismicity, a design basis acceleration of a combination of

Seismic loading | 127

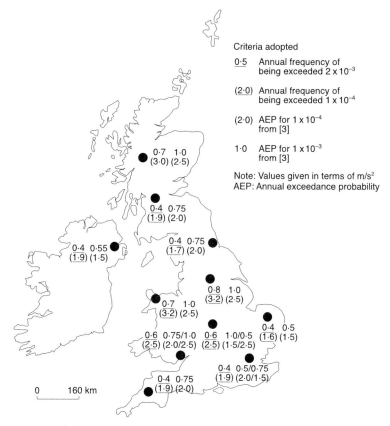

Fig. 9.4. Peak horizontal bedrock acceleration: regional variation in the UK — data from Ove Arup & Partners (underlined) compared with reference [3] (not underlined)

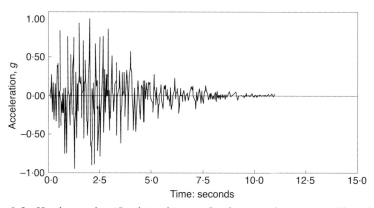

Fig. 9.5. Hard ground artificial time history: first horizontal component (from [4])

0·1g horizontally and 0·07g vertically, and a maximum credible earthquake (MCE) of 0·25g horizontally *or* 0·17g vertically were selected.

The conclusion of what has been said so far is that seismic risk is relevant to dams in the UK. It can be *reasonably* quantified (and engineers with a statutory responsibility for the safety of reservoirs should satisfy themselves) that a dam subject to the Reservoirs Act would not be subject to catastrophic damage in the event of any earthquake which could possibly affect the site.

The Engineering Guide to Seismic Risk to Dams in the United Kingdom

This guide to seismic risk to dams in the UK [4], sponsored by the Department of the Environment, was published in 1991, following comments from dam owners, inspecting engineers, seismic specialists and other interested parties. It has since become quite widely known, discussed and practiced. The guide is very much a guide and not a code of practice and it does not restrict the discretion of the engineer appointed under the Reservoirs Act in deciding criteria to be adopted, techniques to be used and standards to be met. There are limitations to the validity of the guide set by the imprecise state of the art and the lack of records of strong seismic motions in the UK. The author does, however, commend the guide to all those interested and involved in the subject of the safety of dams. Although its authors aimed to produce a concise and readily usable document, it is quite comprehensive. Due to the lack of data on strong seismic motions in the UK, the guide was compiled with the aid of extensive reference to international experience and practice. ICOLD Bulletin 72, entitled *Selecting Seismic Parameters for Large Dams* [6], was given special attention and the guide proposed that dams in the UK should be classified into four categories (I to IV, in ascending order of severity) on the basis of risk classification recommended in that Bulletin. Tables T1 and T2 (reproduced here as Tables 9.2 and 9.3) show the method of risk classification weighted according to capacity of reservoir, height of dam, number of persons at risk and potential downstream damage. An application note to the guide, funded by the DETR [7], which provided some clarifications, was published by Thomas Telford Ltd for The Institution of Civil Engineers in 1998. This note includes results of a study by R M W Musson and P W Winter (*Seismic Hazard of the UK*, Report for DTI by AEA Technology, Warrington, 1996) [8].

Appendix B of the guide [4] gives examples of the behaviour of dams during seismic events worldwide and Appendix C comments briefly on examples of international seismic regulations and criteria. The guide is

Table 9.2. Classification factors (Table T1 from [6]).

Classification factor	4	3	2	1
Capacity: 10^6 m^3	>120 (6)	120–1 (4)	1–0·1 (2)	<0·1 (0)
Height: m	>45 (6)	45–30 (4)	30–15 (2)	<15 (0)
Evacuation requirements: No. of persons	>1000 (12)	1000–100 (8)	100–1 (4)	None (0)
Potential downstream damage	High (12)	Moderate (8)	Low (4)	None (0)

The weighting points of each of the four classification factors shown in parentheses in Table T1 are summed to provide the total classification factor, i.e.

Total classification factor = factor (capacity) + factor (height) + factor (evacuation requirements) + factor (potential downstream damage)

Source: ICOLD Bulletin 72 [6]

Table 9.3. Dam category (Table T2 from [6])

Total classification factor	Dam category
(0–6)	I
(7–18)	II
(19–30)	III
(31–36)	IV

Source: ICOLD Bulletin 72 [6]

therefore of interest to those working on seismic risk to dams in an international context. As one of its authors, the present Author hopes that it will facilitate work on future guides for other countries. The guide deals with the subjects of:

- reservoir safety
- earthquakes in the United Kingdom
- earthquake resistance standards

and for both embankment dams and concrete and masonry dams separately:

- behaviour under seismic loading
- methods of seismic analysis
- seismic safety evaluation.

A much needed glossary and a list of 189 references are also provided.

Perhaps one of the most important features of the guide is Fig. 9.6, which shows an attempt to prescribe three seismic zones throughout the UK:

130 | *Reservoir engineering*

The Channel Isles, not shown, are in Zone A

Fig. 9.6. Seismicity levels in the UK

- Zone A — represents a relatively high number of events, especially larger ones (i.e. with $M_L > 4.5$).
- Zone B — moderate chance of local earthquakes but larger events rare.
- Zone C — few or no recorded earthquakes but some events possible (e.g. Colchester 1884).

Table 9.4. Return period and peak ground accelerations for safety evaluation earthquake (Table T3)

Dam category	Return period: years	Peak ground acceleration, g		
		Zone A	Zone B	Zone C
IV[a]	30 000	0·375	0·30	0·25
III	10 000	0·25	0·20	0·15
II	3 000	0·15	0·125	0·10
I[b]	1 000	0·10	0·075	0·05

[a] In many category IV situations it may be considered desirable to use the maximum credible earthquake calculated from a regional geological and seismological survey
[b] For category I situations seismic safety evaluation is not generally considered to be necessary

The boundaries shown on the map have been drawn subjectively rather than statistically but they have been drawn on the basis of a very high level of experience. By relating the zones to the three levels of seismicity adopted by Irving (Fig. 9.3), in the absence of site-specific data, the guide proposes (for each category of dam) the return period for the safety evaluation earthquake and its horizontal peak ground acceleration considered appropriate, depending on in which of the three zones the dam is situated, as shown in Table 9.2.

Information on a sample of UK dams suggests that only 2·4% or so are likely to fall into Category IV and just over 20% into Category III. The UK guide suggests that dynamic analysis is necessary for dams in Category IV and for concrete or masonry dams over 15 m high in Category III, whereas lower levels of safety evaluation are suggested in cases posing lesser hazards.

It will be as well to gain more experience of the use of the guide in practice but meanwhile it is already encouraging the use of more rational methods in evaluating the safety of dams against more realistic estimates of seismic risk. The guide should therefore be effective in reducing the potential costs of both over- and under-design.

Peak ground accelerations for design in relation to factors of safety and accelerations liable to be suffered

The ultimate stability of a dam depends on all relevant factors, each of which affects the factor of safety or degree of loading which could cause failure. These are:

- geometry of the dam including galleries and shafts (discontinuities)
- any existing cracking or shearing
- reservoir and tail water levels
- uplift (as affected by drainage) and pore pressures

- stress due to self-weight (and due to post-tensioning or cable anchorages if any)
- thermal stresses (including built-in stresses) in concrete dams
- stresses due to geophysical movements of the valley, other than due to the earthquake
- geological conditions in the foundations
- stresses due to a seismic event.

Only the last directly relates to a possible earthquake. The ultimate factor of safety of a dam depends on the reliability with which these parameters can be estimated and with which the water loads and the strengths of materials in the dam and its foundations can be known. Both the ICE *Floods and Reservoir Safety* (see reference [8] in Chapter 6) and reference [3] adopt the axiom also postulated in ICOLD Bulletin 72 [6], that the conditions applied for safety evaluation of a dam should be made increasingly severe according to the severity of the hazard posed by the dam to life and property. The scales of increases in severity of floods and earthquake loading are proposed rather arbitrarily (but reasonably in accordance with international practice) on the principle that the investment in avoidance of failure should be at least of the order of magnitude of the present value of damage which could otherwise be foreseen, as a result of design loadings being exceeded during the intended life of the dam. Table 9.4 shows the allowances made on account of this axiom.

ICOLD Bulletin 72 indicates that for dams whose failure would present a serious social hazard, the maximum design earthquake will normally be characterized by the maximum credible earthquake (MCE), that is, that earthquake which appears possible for the geological conditions and would cause the most severe level of ground motion at the site. Attempts have been and are being made to assign probabilities to the MCE and the maximum possible/probable flood but those attempts have to postulate a reducing gradient in the probability curves as they approach very low probabilities, so that the curves can asymptote to a discrete limiting maximum value appropriate to the site. While limiting values of earthquake magnitude can be conceived, it is necessary to appreciate that the energy released by earthquakes increases many times rather non-linearly for each unit increase on the Richter scale. The largest earthquakes recorded in the world have had a magnitude of about $8 \cdot 9 M_L$. There are scales of magnitude other than the Richter or local magnitude, which are described in the glossary of the engineering guide [4]. In some records and published papers, no distinction has been made. This is, however, not quite as negative as it sounds, because the difference between the scales is quite likely to be within the margin of error of the recording of the magnitude. The Author believes that engineers are more accustomed to thinking in terms of Richter (local magnitude) than of the other scales.

There are also a number of scales of intensity, which are rather subjective scales of the effects of an earthquake on people, on man-made structures or on the surface of the earth, of which probably the scale most familiar to engineers is the Modified Mercalli (MM). The most severe intensity on the scale is XII. Obviously, for a given earthquake, the intensity varies according to ground conditions, but it does not necessarily reduce with distance from the epicentre of the earthquake, because vibrations at depth may be amplified through overlying deposits before they reach the surface, as in the case of the catastrophic earthquake which hit Mexico City some years ago. It is only during the last 60 years that instrumental monitoring of earthquakes has come into use and only in the second half of the twentieth century that monitoring instruments have reached a precision and coverage sufficient to begin to obtain a comprehensive assessment worldwide. Before this, all historical records have had to be based on subjective observations recorded in writing, from which intensities at various locations could be assessed. From these, the epicentres and magnitudes of earthquakes have been deduced. To deduce the focal depth of the earthquakes has always been more difficult and hence subject to greater error. Even using instruments, it is still not easy. However, it must be emphasized that historical observations depended not only on there being people in the locality who felt the effects of an earthquake, but on their experiences being somehow recorded in a retrievable way. This would be less likely in regions with a very sparse population.

The guide to seismic risk [4] draws attention to the United States Bureau of Reclamation's Monograph No. 19 for concrete arch and gravity dams [9], which recommends design for 'unusual loading combinations', 'usual loading combinations' and 'extreme loading combinations'. Only the latter (extreme loading combinations) includes earthquake loading and that recommended is the MCE. The guide to seismic risk proposes factors of safety based on those given in Monograph 19. As an illustration, the minimum shear-friction factor within the dam or at the concrete–rock or masonry–rock contact should be 3·0 for 'usual', 2·0 for 'unusual' and greater than 1·0 for the 'extreme loading combinations'. If conditions resulting in a factor of safety of 1·0 or less implied the imminence of catastrophic failure of a dam, 1·0 or less would not be acceptable, because a factor of safety must contain reasonable provision for uncertainties, margins of error and the unforeseen. Furthermore, if the maximum design earthquake or safety evaluation earthquake is less than the maximum credible (the MCE), it is implicit that there is a risk, however small, of the occurrence of more severe earthquake loading, with a resulting reduction in the factor of safety, should it occur. Much depends on the definition of 'factor of safety': if, in an embankment dam, a factor of safety >1·0 is taken to imply that there is no movement along a slip plane, a factor of safety <1·0 is acceptable, provided that deformations are sufficiently small to pose no threat of the dam being breached and provided that there is no potential for liquefaction in the dam or its foundations.

The UK guide includes a review of simplified and empirical deformation relationships for embankment dams. In 1988, Ambraseys and Menu [10] developed valuable relationships for earthquake-induced ground displacements, U in cm, for a range of probabilities, in terms of the ratio of k_c/k_m where k_c is the critical seismic coefficient which would result in a factor of safety (F) of 1·0 when applied horizontally and k_m is the seismic coefficient applicable, applied horizontally. The relationships are shown in Fig. 9.7, reproduced from the paper by Ambraseys and Menu [10]. The calculation of k_c is readily obtained when standard software is applied to undertake pseudo-static analysis of embankment dams using the method of Sarma. Figure 9.8 is derived from Fig. 9.7 and shows the probability of being exceeded of upper bound earthquake-induced ground displacements U of 0·1 cm, 10 cm and 50 cm against the ratio of the ground acceleration which occurs to the maximum ground acceleration which would cause no displacement. The ratio is 1·8 for a maximum displacement of 10 cm with only 1% probability of being exceeded, which the Author suggests might be a generally acceptable safe limit of displacement, thus tolerating accelerations as much as 1·8 times those for a factor of safety of 1·0 (i.e. zero displacement). It may be noted that the probability of exceedance of the displacement would be much less than the probability of occurrence of the earthquake. An example of the implication of this thinking is that if a 10·0 cm displacement of an embankment dam were considered tolerable under the highest peak ground acceleration (for safety evaluation) of 0·375g proposed in the UK guide [4], the corresponding peak ground acceleration for a factor of safety of 1·0 and no displacement could be expected to be about 0·2g with a 1% probability of being exceeded. The dam could therefore reasonably be designed for an acceleration of only 0·2g with a factor of safety of 0·99. Of course, it might be found in particular cases that the safe limit of displacement were thought to be greater or less than 10·0 cm.

In an article in *Water Power and Dam Construction International* [11] of March 1993, Professor R. Lafitte, Chairman of the ICOLD Committee on Dam Safety, concluded, *inter alia*, that the concept of risk

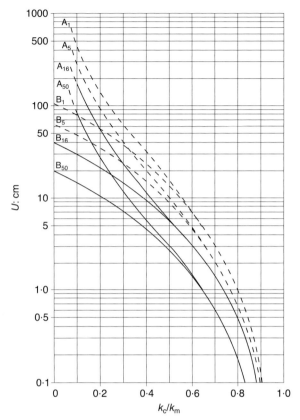

Fig. 9.7. Earthquake-induced ground displacements. (After Ambraseys and Menu [10]). The figure shows predicted values of maximum unsymmetrical (A) and symmetrical (B) displacements for probabilities of exceedance of 1, 5, 16 and 50%

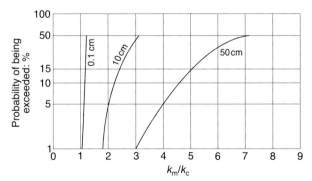

Fig. 9.8. Upper bound earthquake-induced ground displacements

acceptance at present, leads to the recommendation that the dam should be designed so that the risk of death is no higher than 1 in 1 million per year person. Professor Lafitte has also stated as an axiom that it would not be acceptable on moral grounds to put a monetary value on human life in calculations of damage. This sort of thinking encourages the adoption of extreme loading which will never be exceeded as far as can be foreseen. This, in effect, would rule out foreseeable failure and potential resulting death. The possibly expensive consequences of requiring dams to such a standard of unlimited safety have drawn the Author to the rationale that a lower standard could be accepted if a system were to be established for warning and evacuating people in the area liable to be affected by an impending failure of a dam, in a way which would virtually guarantee that no lives could be lost (an Emergency Preparedness Plan or EPP). The EPP would have to be put into effect should seismic loading occur greater than a threshold above which failure of the dam could be expected to occur.

The same approach might be adopted in responding to the impact of extreme floods, where failure of a dam is also the hazard to be guarded against. An amplified discussion of this approach appears in Gosschalk *et al.*, 1994 [12].

It might be thought that the publication of EPPs would cause serious social and economic difficulties. However, evidence to the contrary exists. For example, British Columbia Hydro in Canada have prepared EPPs for its dams and tests the EPPs on a regular basis. An operational exercise reported by N. M. Nielsen in *Water Power and Dam Construction* in March 1993 [13] involved a simulated earthquake-triggered dam breach with B C Hydro's Corporate Emergency Centre and 23 agencies participating. Nielsen reported that these exercises were well accepted by the public and agencies. Yorkshire Water has prepared such contingency plans for all reservoirs for which it is responsible under the Reservoirs Act. More than half of 34 countries included in a survey require hazard mapping and emergency plans, at least for major dams. Hazard mapping implies estimating and mapping the extent and depth of flooding which would be caused in the event of breaching of a dam.

Case histories of seismic incidents

Very refined techniques can be applied to the seismic analysis of earthfill, concrete and masonry dams. Such work is valid, however, only so far as it describes the behaviour of dams under real seismic loading. Therefore case histories of dams which have suffered damage and also of dams which have withstood significant shaking without damage will now be considered. The Author is grateful to his colleague Jonathan L. Hinks for compiling most of the data under review here [14].

Seismic loading | 137

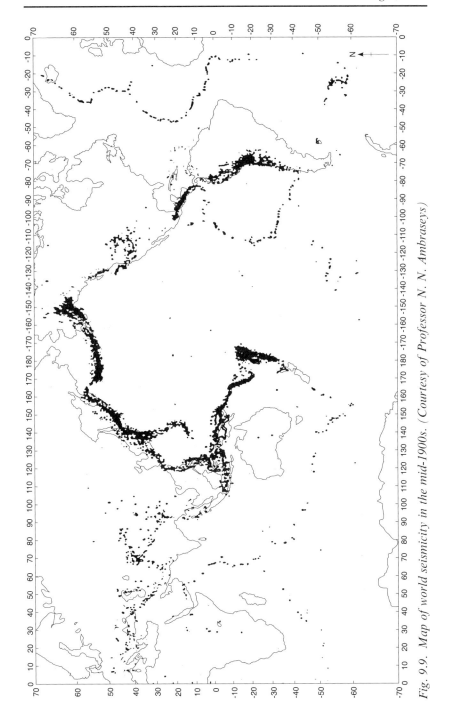

Fig. 9.9. Map of world seismicity in the mid-1900s. (Courtesy of Professor N. N. Ambraseys)

Figure 9.9 shows a map of world seismicity in the mid-1900s reported by Professor N. N. Ambraseys. The concentrations of events around what are believed to be the edges of global tectonic plate movements are particularly noticeable around the periphery of the Pacific Ocean. Reference will be made to events affecting dams in the western United States, Chile, Mexico, Iran, Greece, Japan, Italy and China. Cases of damage in relatively quiet seismic areas such as western India and the UK will also be noted.

There were significant earthquakes affecting dams in Chile in 1928, 1943, 1965 and 1985 [15–18].

On 1 October 1928, the Barahona tailings dam at the El Teniente Copper Mine failed a few minutes after a large earthquake which lasted 1 min and 40 s. At the time of the failure, the dam contained 27 Mtonnes of material. The dam was 63 m high. The failure seems to have been initiated by liquefaction of the core and the consequent inward slide of the embankment. The result was that 4 Mtonnes of material flowed into the valley, killing 54 people.

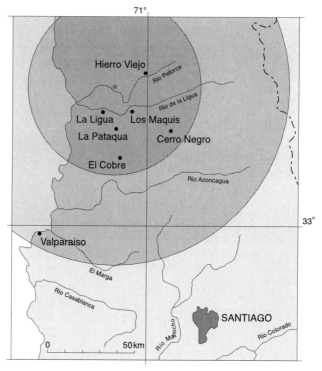

Fig. 9.10. Location map of La Ligua earthquake, Chile, 1965. (After Dolary and Alvarez)

On 28 March 1965, there was an event of Richter magnitude 7 to $7\frac{1}{4}$ with epicentre near the town of La Ligua shown on the map in Fig. 9.10. The depth of its focus (focal depth) was about 61 km. The worst damage was at El Cobre tailings dams about 40 km south of the epicentre. Two of the three dams at El Cobre were almost completely destroyed and more than 2 Mtonnes of tailings flowed into the valley, travelling 12 km in a few minutes, destroying part of the town of El Cobre and killing more than 200 people.

A maximum horizontal acceleration of $0.18g$ was recorded in Santiago some way to the south, so the peak acceleration at the site was probably higher than this.

A third dam at El Cobre suffered only local slides. This dam was no longer in use, which suggests that it might have been drier than the others. There were also failures at the Hierro Vejo dam, one of the dams at Los Maquis (where the core is reported to have liquefied) and at La Pataqua Dam. There were four tailings dams at Cerro Negro, of which only one was in use. The one in use failed spectacularly, whereas the others suffered only fracturing and small slides.

In the 1985 earthquake, magnitude 7·8, two dams collapsed and 16 suffered major damage, as shown on the map on Fig. 9.11. A further 16 suffered only minor damage although exposed to similar levels of shaking. The two that collapsed were the La Marquesa and La Palma de Quilpue earthfill dams, each 10 m high. The peak horizontal ground acceleration at La Marquesa was estimated at $0.6g$ and that at La Palma as $0.46g$.

La Marquesa Dam experienced major sliding of both upstream and downstream slopes, resulting in about 2 m loss of freeboard. De Alba and Bolton Seed [19] postulated that the contact layer of silty sand on which the main body of the embankment was constructed, liquefied soon after the onset of the strong motions.

At La Palma Dam, major sliding took place on the upstream slope. A major crack developed along the crest with a maximum width of 1·2 m. The upstream side of the crack settled more than 800 mm relative to the downstream side. As at La Marquesa, the slope failures have been attributed to liquefaction of loose sand layers near the base of the embankment.

These Chilean earthquakes seriously damaged a large number of dams and many lives were lost. They point to the care needed to resolve the potential for liquefaction of earthfill dams and also of the foundation material. Some conclusions from the work of Professor Retamal [17] are as follows:

- a minimum intensity of about $0.1g$ is required to produce damage

140 | *Reservoir engineering*

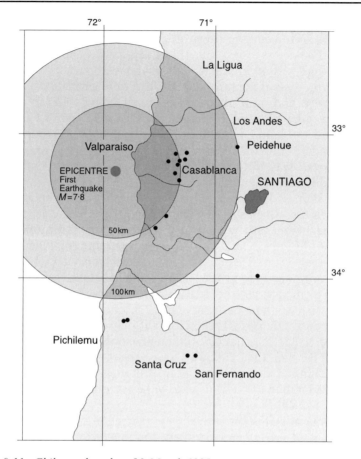

Fig. 9.11. Chile earthquake of 3 March 1985

- virtually no damage occurs at distances over 180 km from the epicentre
- a correlation of damage occurring can be detected with the orientation of the longitudinal axis of the dam.

These points are useful as guides but cannot be taken as absolute rules. There are a considerable number of different attenuation formulae in use, all of which have their own conditions of applicability. However, when plotted on log paper, most look similar to that of Joyner and Boore for earthquakes in the western United States (Fig. 9.12). The curve suggests that there is probably not much attenuation for distances of less than about 10 km from the epicentre but that, even for quite large events, the peak ground acceleration will be very small at a distance of 180 km from the epicentre. Professor

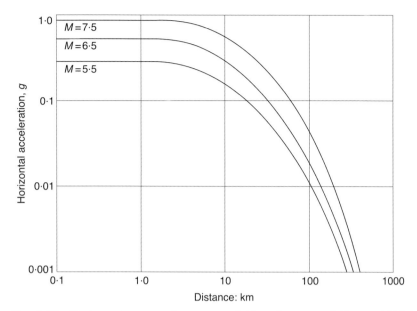

Fig. 9.12. *Horizontal attenuation curves. (After Joyner and Boose, 1981.) Note: 183 recordings of peak acceleration from 24 shallow earthquakes in Western North America*

Retamal's final point about the orientation of the dams is interesting and confirms that it is impossible to reduce seismic inputs to a single parameter: the peak horizontal ground acceleration is obviously useful, but other parameters such as the orientation of the dam, the spectral intensity and duration of shaking are also important and need to be considered. ICOLD Bulletin 72, for example, points out that an event of large magnitude tends to be more damaging because of the longer duration of shaking. Tables 9.5 and 9.6 summarize damage to some earthfill dams including those which have been mentioned. Similar lists are shown for rockfill and concrete dams in Tables 9.7 and 9.8.

The 1906 San Francisco earthquake had a Richter magnitude of $8\frac{1}{4}$ when there were 33 earth dams within 56 km of the fault and 15 within 8 km. It seems likely that all these dams were subjected to ground motions having peak ground accelerations greater than $0.25g$ and that those within 8 km probably experienced accelerations greater than about $0.6g$. Yet none of these old dams suffered any significant damage. In his 1979 Rankin lecture [20], H. Bolton Seed pointed out that the slopes were fairly steep (typically 1:2 to 1:3) and that the dams had generally been compacted by moving livestock or by

Table 9.5. Damage to some earthfill dams, 1

Dam	Height: m	Location	Date	Damage
Barahona Tailings Dam	63	Chile	1928	Catastrophic failure: 54 killed
El Cobre Tailings Dams	32–35 19	Chile	1965	2 dams destroyed: 200 killed
Cerro Negro Tailings		Chile	1965	1 dam failed
La Marquesa and La Palma	10 10	Chile	1985	2 dams failed 16 dams — major damage 16 dams — slight damage
San Andreas Dam	32	California	1906	Longitudinal cracking. Transverse cracking at abutments
Upper Crystal Springs	26	California	1906	2·4 m offset movement in dam
Sheffield Dam	8	California	1925	Complete failure
Hebgen Dam	35	Montana	1959	Dam settled 1·2 m. Seismic seiches

teams and wagons. He pointed out, however, that they were all constructed of clayey soils on rock or clayey foundations. Two dams were built largely of sand but this was apparently not saturated.

The Santa Barbara earthquake of 29 June 1925 caused the complete failure of the 8 m high Sheffield Dam in California. The water level was 4·5 m below the crest at the time. The earthquake had a Richter magnitude of 6·3 and the epicentre was about 11 km from the dam site. Both the dam and its foundation were predominantly silty sand. A length of 100 m of the dam moved bodily 30 m downstream, releasing 200 000 m^3 of water. The maximum ground acceleration has been estimated at 0·15g. Liquefaction has been suggested as the cause of failure although this explanation is not universally accepted. Had the reservoir been full at the time of the earthquake, the situation would obviously have been much more serious.

On 17 August 1959 the Hebgen Dam (35 m high) in Montana was subjected to an earthquake quoted as having a Richter magnitude of 7·1 (although others have given the magnitude as 7·5 to 7·8). The embankment was built as rolled fill, from gravelly clay of medium plasticity. It had a thick central concrete core. One of the main faults passed along the shore of the reservoir about 215 m from the dam. Along the fault, there was a maximum vertical displacement of the ground surface of 4·5 to 5·5 m.

Table 9.6. Damage to some earthfill dams, 2

Dam	Height: m	Location	Date	Damage
Lower San Fernando	40	California	1971	Crest settled 8·5 m
Upper San Fernando	24	California	1971	Crest settled by 900 mm. Moved downstream 1·5 m
Paiho Main Dam	66	China	1976	Large slide: 330 dams damaged
Douhe Dam	22	China	1976	Longitudinal cracking
Masiway Dam	25	Philippines	1990	1·0 m crest settlement. Longitudinal cracking
Ono Dam	37	Japan	1923	250 mm crest settlement, deep fissure adjacent to core wall
74 embankments	1·5 to 18	USA	1939	Cracks, leakage
Chatsworth	11	USA	1930	Cracks, leakage
Earlsburn	6	Scotland	1839	Complete failure
Lower Van Norman	45	USA	1972	Slump of u/s face*
Nihon-kai-Chubu Earthquake	—	Japan	1983	145 dams failed or suffered heavy damage
Matahina Dam	86	New Zealand	1987	Up to 800 mm settlement

* Dam failed due to liquefaction in the body of the dam

As shown on Fig. 9.13, the dam did not fail although it settled by up to 1·2 m on either side of the core wall, which was left standing proud of the dam. The core wall was itself somewhat cracked and the spillway was damaged. An interesting feature was the seismic seiche set up in the reservoir. The caretaker observed a wave striking the dam, so that about 900 mm depth of water poured over the crest for 10 min. The wave then receded but returned after a further 10 min. This continued with the dam being overtopped at least four times. The length of the reservoir was about 27 km.

The San Fernando earthquake in California on 9 February 1971 had a Richter magnitude of 6·6. The relatively modern earth dams in the area performed well but the Lower San Fernando Dam (40 m high, Fig. 9.14), which was constructed of hydraulic fill, was severely damaged by the motion which was estimated to have had a peak ground acceleration of 0·55 to 0·6g.

The earthquake caused a major slide in the upstream shoulder of the dam, taking out the crest and the upper 9 m of the downstream slope. Failure seems to have been initiated by liquefaction of hydraulic sand fill, in the lower section of the upstream shoulder. Figure 9.15, showing

Table 9.7. Damage to some rockfill dams

Dam	Height: m	Location	Date	Damage
Cogoti Dam	85	Chile	1943	600 mm settlement
Minase Dam	67	Japan	1964	61 mm settlement. Minor joint damage to concrete face
La Calera Dam	30	Mexico	1964	Overtopped by waves 2·5 m high. 120 m long slip
Oroville Dam	230	California	1975	10 mm settlement 150 mm downstream movement
El Infiernillo	148	Mexico	1979	130 mm settlement
La Villita	60	Mexico	1979	50 mm settlement
Austria Dam	56	California	1989	300 mm settlement Deep cracks 30 mm settlement
Miboro Dam	131	Japan	1961	50 mm downstream movement
Malpaso Dam	70	Peru	1938	76 mm settlement 51 mm downstream movement
Malpaso Dam	70	Peru	1958	32 mm settlement 58 mm downstream movement

a reconstruction of the embankment, is due to Professor Bolton Seed and shows the area where liquefaction is thought to have taken place.

It is very fortunate that the water level in the reservoir was about 10 m below the crest at the time of the earthquake. After sliding, only about 1·5 m of badly cracked material remained above water level. Eighty thousand people living downstream of the dam were evacuated until the water level was reduced.

As shown in Fig. 9.16, the Upper San Fernando Dam (24 m high) was also affected in the same earthquake. The crest settled by 900 mm and moved downstream 1·5 m.

As shown in Table 9.6, in the 1976 Tangshan earthquake in China, some 330 dams were damaged including the Paiho Main Dam (66 m high). The earthquake had a magnitude of 7·8. The Douhe Dam (22 m high) was one of the others damaged in the same event. There was extensive longitudinal cracking, crest settlement and heaving of the toes of the embankment. The damage was attributed primarily to liquefaction of the saturated silts in the foundation.

Damage to the Masiway Dam (25 m high) in the Philippines (Fig. 9.17), in the 7·7 magnitude earthquake in 1990, has been described by Derek Knight in the November 1991 issue of *Dams and Reservoirs*, the journal of the British Dam Society. As can be seen in Fig. 9.17, the crest cracked longitudinally and settled by up to 1 m.

Table 9.8. Earthquake effects on some concrete dams

Dam	Height: m	Type	Location	Date	Damage
Koyna	103	Concrete gravity	India	1967	Major cracking
Sefid Rud	106	Buttress	Iran	1990	Major cracking
Pacoima	113	Arch	California	1971	Cracking at left abutment
Lower Crystal Springs	47	Curved gravity	California	1906	No damage
Blackbrook	29	Concrete and masonry gravity dam	UK	1957	Copings displaced. Cracking
Hsingfengkiang	105	Buttress	China	1962	Major crack
Honen-Ike	30	Multiple arch	Japan	1946	Crack in arch near buttress
Ambiesta	59	Arch	Italy	1976	No damage
Mama di Sauris	136	Arch	Italy	1976	No damage
Shenwao	52·6	Concrete gravity	China	1975	Cracking
Redflag	35	Masonry gravity	China	1970	Cracking, leakage
Rapel	110	Arch	Chile	1985	Damage to spillway and intake tower

A comprehensive survey was carried out of 74 embankments severely damaged during the magnitude 6·6 Ojika earthquake in Japan in 1939. The heights of the embankments ranged from 1·5 to 18 m and the intensity of shaking was approximately 0·3 to 0·4g. Twelve of the dams failed completely and there were 40 reported cases of slope failures. The conclusions of the survey are summarized as follows:

• the majority of damaged and failed embankments consisted of sandy soils
• no complete failures occurred in embankments constructed of clay soils
• very few cases of dam failures occurred during the earthquake; most failed either a few hours or up to 24 hours after the earthquake.

The last observation points out possible effects of the migration of pore pressures generated by the earthquake not only during the period of shaking itself but also for hours or maybe even days following the event.

In the Nihon-Kai-Chubu earthquake in Japan in 1983, 145 dams suffered failure or its equivalent, that is, 13% of the dams in the

146 | Reservoir engineering

Fig. 9.13. Hegben Dam, Montana, 1959

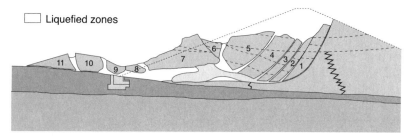

Fig. 9.14. Lower San Fernando Dam, California: cross section through embankment after 1971 earthquake

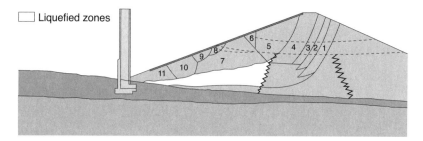

Fig. 9.15. Lower San Fernando Dam: reconstructed cross section

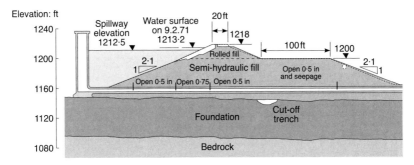

Fig. 9.16. Cross section through Upper San Fernando Dam, 1971. (After Los Angeles Dept. of Water and Power)

area. Peak ground accelerations were thought to be between 0·08 and 0·25g (incidentally accelerations which many think could occur in the UK). The following points were noted.

- Flatter slopes were more prone to damage. (Pseudo-static analyses suggest that for dams with the same factor of safety under static conditions, the dam with the flatter slope will be more vulnerable to an earthquake.)
- Sand embankments were the most vulnerable, being 2·5 times more likely to be heavily damaged than clay embankments.
- Vulnerability was greater with low reservoir levels

So far, only failures in earth dams have been described, many of which can probably be attributed to liquefaction either of the foundation or of the dam itself. There are, of course, other types of dam and other types of failure, although the prominence given to liquefaction is justified by the large proportion of failures of earthfill dams due to this cause.

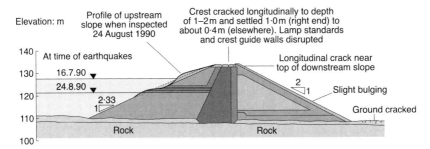

Fig. 9.17. Earthquake damage to Masiway Dam, Philippines, 1990

Before turning to rockfill and concrete dams, the case of the Earlsburn Reservoir (6 m high dam) near Stirling in Scotland will be noted. This was washed away on 23 October 1839, although the *New Statistical Account of Scotland* recorded that 'on the evening of the 25 October 1839 the waters burst forth and did considerable damage but whether it was owing to the earthquake which was distinctly felt in many parts of the parish that night has not been clearly ascertained'. Dr R. M. W. Musson of the British Geological Survey, Edinburgh, has tried to unravel what actually happened and reported his results in the May 1991 issue of *Dams and Reservoirs*: the earthquake had a magnitude of about 4·8, and was felt over a wide area; the dam was an embankment of peat and earth with a narrow central core of silty clay. The core extended down to rock but most of the dam was founded only on peat. The failure appears to have taken place about 8 hours after the earthquake. Dr Musson concluded that the earthquake probably triggered the dam burst although it may not have been the principal cause of failure. No lives were lost although there was quite a lot of damage to livestock. There was one benefit from the failure, in that a large number of trout from the reservoir were left by the flood in such numbers on the fields that the local people collected them in basketfuls.

On the list of damaged rockfill dams (Table 9.7), Cogoti Dam, 85 m high in Chile, is a concrete faced dam. On 6 April 1943, it was affected by a 7·9 magnitude event with epicentre 89 km from the dam. The peak ground acceleration at the site was estimated as $0·19g$. The crest of the dam settled about 600 mm. Cogoti Dam was again damaged in the earthquake of 8 July 1971, when a longitudinal crack appeared in the crest. The peak ground acceleration on that occasion was estimated at only $0·05g$.

Without going through all the dams listed, it can be pointed out that damage is most typically described as crest settlement, which is sometimes combined with downstream movement. None of these rockfill dams suffered catastrophic failure. An interesting feature at El Infiernillo and La Villita [21] is that both dams had accelerograms on the crest and on rock near the dam; the acceleration on the crest at La Villita was between 9 and 22 times as great as that measured on rock at the right abutment, depending on whether the transverse, longitudinal or vertical acceleration is considered. At El Infiernillo, which is 2·5 times the height of La Villita, amplification was between only 2·7 and 4·8. It is probably of significance in this respect that El Infiernillo dam is founded on rock whereas La Villita is founded on deep alluvial deposits [21].

An interesting point to note with the Malpaso Dam (70 m high) in Peru, is that the crest settled 76 mm in the 1938 earthquake even though the intensity was only VII on the Modified Mercalli (MM) scale. (Ground acceleration was less than $0.05g$.) The 1958 earthquake was much bigger but caused settlement of only 32 mm.

The question of the performance of concrete-faced rockfill dams was raised at a half-day meeting at the ICE on 29 January 1992. It is generally thought that a concrete-faced type of rockfill dam is a good choice in a seismic area (the fill is separated from saturation by the reservoir throughout its cross section) but there is some concern about possible damage to the concrete membrane (which, however, is more accessible for repairs if required, than would be a membrane within the dam). At the meeting, Dr Penman pointed out that despite the settlement at the Cogoti dam, the membrane remained intact, probably because it was constructed in bays with copper water stops (as is usually the case with this type of dam).

In his 1990 Binnie lecture, however, Dr Paul Back noted the poor behaviour of concrete slabs on earth dams in the Philippines and said that this raised questions as to the seismic performance of dams which depend on concrete faces as the main water barrier. (The Author would say only that the use of concrete faces on earth dams needs special care and appropriate techniques, to allow for differential settlement.)

Turning to concrete dams, the damage to Koyna Dam (103 m high) in India, shown in Fig. 9.18, is a classic case, which has been very fully documented and extensively studied. The area in which the dam is located was formerly considered to be seismically stable and it is now generally believed that the earthquake of 10 December 1967, which had a magnitude of about 6·4, was induced by the filling of the reservoir. The earthquake was preceded by hundreds of similar earthquakes which began to occur near the dam after the reservoir behind the dam started to fill in 1962. Apart from the damage to the dam, about 200 people were killed by the earthquake.

Peak accelerations of $0.63g$ longitudinally, $0.49g$ transversely and $0.34g$ vertically were recorded by seismographs located about 20 m below the crest. Cracking at Koyna, as shown in Fig. 9.19, took place near the change of slope on the downstream face. Significantly, the cracking stopped short of the overflow (spillway) monolith sections, where inertial forces would have been much less.

As shown in Fig. 9.19, the dam has a rather unusual cross section because it had originally been planned to construct the dam in two stages. It has been suggested that the extra weight at the top of the dam contributed to the damage. However, Chopra and Chakrabati [22] have commented:

150 | *Reservoir engineering*

Fig. 9.18. Koyna Dam, India

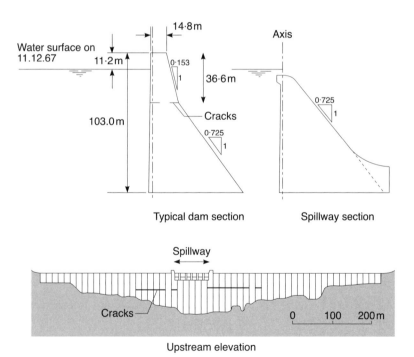

Fig. 9.19. Koyna Dam, typical section

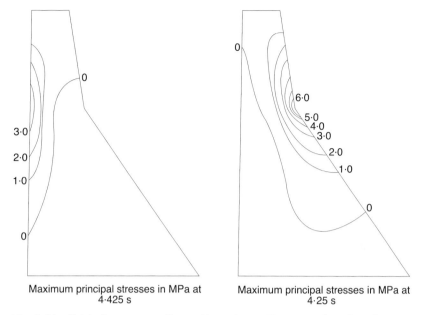

Fig. 9.20. *Critical stresses in Koyna Dam due to Koyna earthquake, 1967*

> Intuitively, the Koyna section would appear to be more vulnerable to earthquake damage than a typical gravity dam section. However the analytical results... do not support such a contention. They demonstrate that comparably large tensile stresses would develop even in typical gravity dam sections.

Peak tensile stresses calculated for Koyna shown in Fig. 9.20 were about 3·4 MPa (500 p.s.i.) on the upstream face and 6·9 MPa (1000 p.s.i.) on the downstream face. It has been shown that the additional tensile stresses caused by hydrodynamic pressures acting on the upstream face of the dam represent roughly 45% of these maximum values. This can be illustrated by reference to some theoretical work by Chopra and Gupta [23] for the Pine Flat Dam. They showed that when foundation flexibility and hydrodynamic effects are taken into account, the calculated stresses are greatly increased (Table 9.9).

Herzog [24] has pointed out that, if the pseudo-static method of analysis is applied to Koyna, then it is necessary to factor up the actual peak ground acceleration from 0·49g to 1·1g to get the observed stresses. It is because the pseudo-static method can underestimate stresses in this way that dynamic analysis has been recommended for UK dams in Category IV and for dams higher than 15 m in Category III.

The Sefid Rud Buttress Dam (106 m high) in Iran (Fig. 9.21), was affected by an earthquake in June 1990 which had a magnitude of

Table 9.9. *Tensile stresses at Pine Flat Dam resulting from Taft horizontal ground motion only*

Case	Foundation rock	Hydrodynamic effects	Maximum tensile stresses: MPa		
			Upstream face	Downstream face	Heel
1	Rigid	Excluded	1·05	1·43	1·77
2	Rigid	Included	1·54	1·75	2·52
3	Flexible	Excluded	0·99	1·09	1·03
4	Flexible	Included	2·05	2·45	2·95

7·3 to 7·7. The earthquake was very destructive and it has been estimated that 40 000 people were killed and more than 100 000 buildings were destroyed or badly damaged. The dam was about 5 km from the epicentre. The nearest accelerograph which was working was about 40 km away and recorded a peak ground acceleration of 0·56g. A

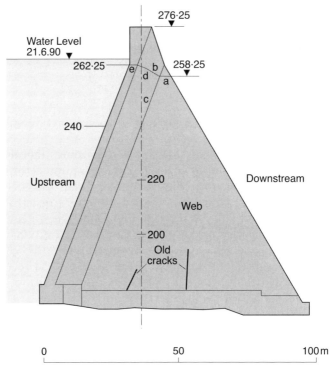

Fig. 9.21. *Sefid Rud Buttress Dam, Iran. (After Ahmadi and Khoshrang)*

recent paper by Ahmadi and Khoshrang [25] suggests that the peak horizontal ground acceleration at the site would have been over $0.7g$, assuming the Gutenberg–Richter formula.

Most of the damage was in the top 25% of the buttresses. Major cracks about 10 mm wide developed along horizontal construction joints, near to the change of slope on the downstream face. Unlike at Koyna, the level of cracks varied from monolith to monolith. At monolith 15 there was a 20 mm displacement of the crest of the dam towards the downstream side, with severe leakage through the crack. There was also severe damage to the parapet which must have experienced very high accelerations as a result of amplification. There was also some relative movement between buttresses. The damage might have been worse had not the dam been apparently strengthened by post-tensioned anchorage blocks at the toes of the buttresses and lateral buttressing between buttresses (not shown in Fig. 9.21).

The Pacoima Dam (113 m high, Fig. 9.22) in California is an arch dam. On 9 February 1971, there was an earthquake of magnitude 6·6 with the epicentre 6·4 km from the dam. The focus was at a depth of about 12·8 km. The thrust fault on which major movement occurred, slopes upwards, passing below Pacoima Dam at a depth of about 4·8 km.

Measurements made after the earthquake showed that the distance between the abutments of the dam had shortened by 24 mm. There was a slight tilting of the dam with the right end dropping 17·3 mm relative to the left end. The only visible damage to the structure was opening of the radial contraction joint between the arch and the left abutment thrust block. The opening was 6·3 mm to 9·7 mm and extended to a depth of 13·7 m.

As shown in Fig. 9.23, there was an accelerograph at the site which measured high-frequency peaks of $1.25g$. However, the accelerograph was located high up on a ridge near the abutment and it is thought that the base rock acceleration was probably 0·6 to $0.8g$ rather than $1.25g$.

Lower Crystals Springs Dam (38·7 m high) (Fig. 9.24) is a curved concrete gravity structure about 24 km south east of the city of San Francisco. The earthquake of 18 April 1906 had a magnitude of 8·3, with the dam only 122 m from the fault. Remarkably, the dam withstood the earthquake without damage although recent analysis suggests a peak tensile stress of 5·7 MPa (830 p.s.i.) at the extreme end of the downstream face. The highest tensile stress within an interior element was calculated as 4·0 MPa (588 p.s.i.) at the top of the downstream face.

Three factors seem to be relevant, illustrated in Fig. 9.25: firstly, the very generous cross section; this was because the dam was designed

Fig. 9.22. Pacoima Arch Dam, California

for further raising which never took place. Secondly the fairly low height of the dam compared to, say, Koyna, and thirdly the high standards of construction. The dam was completed in 1890, but unusually careful attention was paid to quality control in mixing and placing the concrete. For example, it was one of the earliest dams where the water–cement ratio was established by specification.

It is clear that the concrete at Lower Crystal Springs was subjected to high tensile stresses without damage, and this raises the question as to what level of dynamic tensile stresses can be accepted in a concrete dam? The slide shown earlier for Koyna gave peak tensile stresses of 3·4 MPa on the upstream face where serious cracking occurred, so the experience at Lower Crystal Springs cannot be used to justify the use of allowable tensile stresses as high as 4·0 MPa. There is, of

Fig. 9.23. Pacoima Dam, strong motion accelerograph site

course, a fair amount of evidence supporting the use of dynamic tensile strength higher than the static tensile strength of the concrete. In his paper in the August 1980 issue of *Water Power and Dam Construction* [26], Herzog proposed dynamic increase factors of 20% for a frequency of 3 Hz and 50% for a frequency of 10 Hz. Work by Raphael [27] in the United States suggests a figure of 50%. It is probably safe to think of the dynamic tensile strength being perhaps 30% greater than the static tensile strength of the concrete. It must be remembered, however, that

Fig. 9.24. Lower Crystal Springs Dam, California

incipient cracks probably exist in most concrete dams as a result of static thermal loadings (ICOLD Bulletin 52, p. 43 [28]) and also that the lift joints are likely to be the areas of greatest tensile weakness. It is for these reasons that it has been suggested in the UK guide that the assumed dynamic tensile strength at lift joints may be taken as the minimum dynamic strength determined, not exceeding 2·0 MPa.

Figure 9.26 shows Blackbrook Dam (30·5 m high) near Loughborough in the UK. Blackbrook is a concrete gravity dam with brick

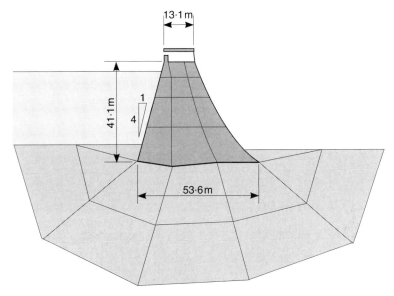

Fig. 9.25. Lower Crystal Springs Dam — cross section. (After Wolff and Orden)

and masonry facing. The earthquake, of local magnitude 5·3 at a focal depth of 7 km, was on 11 February 1957 with the epicentre about 6·5 km north of the structure. The main visible damage was displacement of the 0·75 tonne coping stones on the parapets. There were also six vertical cracks in the masonry arches over the spillway, and cracks in the mortar of the stone facing on the downstream face. On the day of the earthquake, the leakage flow increased to 10·5 l/s. By 6 March, it had reduced to 1·6 l/s. This temporary increase in leakage was thought to be caused by disturbances in the foundation rock rather than by leakage through the dam. Incidentally, similar increases in leakage were noted at the Izvorul Montelui gravity dam in Romania in 1977 and at the 52·6 m high Shenwao concrete gravity dam in China in the Hsicheng earthquake in 1975.

The Hsinfenkiang Dam (105 m high) in China, shown in Fig. 9.27, is a diamond head buttress dam on a granite foundation. It was completed in 1959 having been designed for an earthquake with intensity VI MM. However, there were frequent earthquakes after impounding, so, in 1961, it was strengthened for intensity VIII. These earthquakes are generally believed to have been caused by impounding the reservoir. On 19 March 1962. there was an event of magnitude 6·1 with epicentre about 1 km from the dam. An aftershock of magnitude 4·5 produced a peak acceleration of 0·54g on the crest. A horizontal

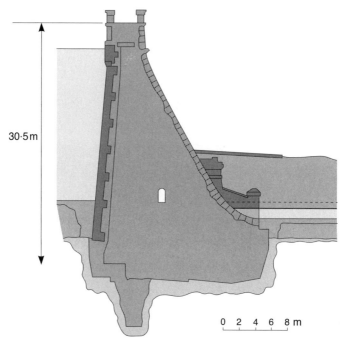

Fig. 9.26. Blackbrook Dam, UK

continuous crack about 82 m in length appeared, about 16 m below the top of the right bank non-overflow monoliths.

A considerable number of other concrete dams have been shaken by earthquakes; quite often these events are thought to have been induced by filling the reservoir: examples are the Hoover Dam (USA), Monteynard (France), Kariba (Zimbabwe), Kurobe (Japan) and Grandval (France). However, none of them suffered any damage. There are many concrete dams which have withstood severe shaking without damage. The Ambiesta arch dam in northern Italy is 59 m high and was only 22 km from the epicentre of the Germona-Friuli earthquake of 6 May 1976 (magnitude 6·5). The earthquake caused 965 deaths and damage estimated at $2·8 billion. A maximum acceleration of 0·33g was measured at the right abutment of the dam. Neither the Ambiesta dam nor 13 other concrete arch dams in the area suffered damage from this event and this includes the 136 m high Maina di Sauris dam located 43 km from the epicentre of the magnitude 6·5 event.

In addition to the dams, it is necessary to consider ancillary structures such as culverts and intake towers. Other structures, such as barrages, should also be considered, of which over 100 were destroyed

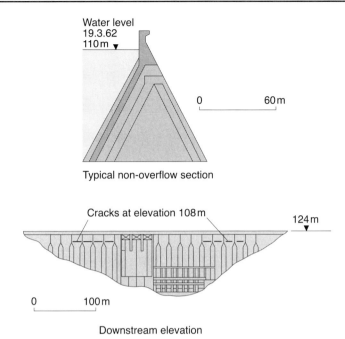

Fig. 9.27. Hsinfenkiang Dam, China — section and elevation

in the 1976 Tangshan earthquake in China. Shen, Huang and Wang [29] believe that this was because they tended to be founded on soft material rather than rock.

Conclusions from case histories on improving the earthquake resistance of dams

(a) The worst damage from earthquakes has often been associated with liquefaction of embankment materials or of the foundations. Sandy materials with particle size lying between 0·05 and 5 mm, and especially between 0·1 and 1 mm, seem to be particularly susceptible to liquefaction. It is believed that soils are not liquefiable if they have more than 20% finer than 0·005 mm particle size and also if the liquid limit (of fine material) is more than 35% and the plasticity index (of fine material) is greater than 10. Thus materials outside these limits should be avoided as far as possible in embankment dams or in their foundations in seismically active areas. The provision of good drainage downstream of the impermeable membrane, and filters to prevent internal erosion due to drainage or leakage, is highly important, to avoid saturation conditions and hence to reduce the potential for liquefaction [30,31].

160 | *Reservoir engineering*

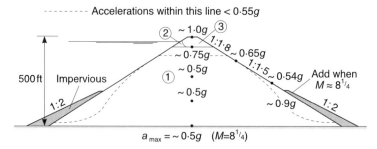

1 Compacted rockfill
2 Rollcrete or soil/cement
3 Compacted clayey sand and gravel

Fig. 9.28. *Conceptual section for 500 ft concrete-faced rockfill dam in highly seismic region*

(b) Material placed as hydraulic fill appears to be particularly vulnerable — possibly on account of its high moisture content and low density. Its use should be avoided in seismically active areas.

(c) Clay dams on clay foundations generally perform well when subjected to earthquakes. However, a number of clay embankments were damaged in the Nihon-Kai-Chubu earthquake in Japan in 1983.

(d) Rockfill dams generally perform well, particularly if well compacted. Figure 9.28 shows a conceptual design of section by Professor H. Bolton Seed and others [32] for a 500 ft high concrete faced rockfill dam in very highly seismic conditions. It can be noted that mass is added by flattening the slopes approaching the crest and by flattening the slopes at the upstream heel and downstream toe.

(e) Concrete dams generally perform well although there are several cases of damage, particularly to buttress dams. Figure 9.29 shows the basic section for a typical UK gravity dam which phases-in the out-of-balance weight at the crest. Resistance to more severe seismic loading can be gained by flattening the downstream slope to $0.75:1.0$, $0.8:1.0$ or even flatter, thus increasing the mass of the dam and, of course, its cost.

(f) Sudden changes in section or alignment should be avoided or relieved by transitions and if necessary, should be reinforced to avoid the concentration of stresses and initiation of cracks. The inclusion of inspection galleries within concrete dams needs particular care in this respect. Figures 17.2 and 17.3 of Chapter 17 illustrate the problem. Rounded sections and/or external reinforcement might be desirable.

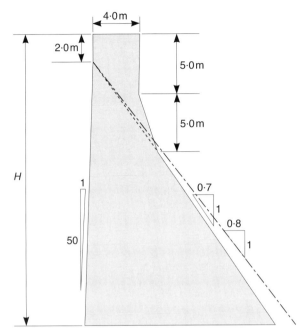

Fig. 9.29. Basic section for typical UK gravity dam (after Daniell et al.)

(g) With buttress dams and hollow gravity dams, to improve earthquake resistance, the provision of transverse stiffeners or diaphragms and/or cable foundation anchorages of buttresses, to increase the rigidity of the structure under dynamic loading, should be considered.

Conclusion

As a brief final conclusion, the author believes that a new era in the assessment of the safety of dams in the event of earthquakes dawned with the adoption of *An Engineering Guide to Seismic Risk to Dams in the UK*, and that the approach recommended is rational, practicable and economic. However, without doubt much more research over a long period of time is needed into the nature and effects of earthquakes. There is indeed doubt whether it will ever be possible to forecast the occurrence and magnitude of earthquakes with the reliability needed, considering the depths below the Earth's surface at which the events occur. Having said that, it is noted that a Greek Professor has claimed a good success rate in predicting earthquakes on the basis of electric signals which, he says, emanate from the Earth preceding earthquakes.

References

[1] Long, R.E. *A ground motion probability analysis for Britain based on macroseismic earthquake data*. Proceedings of Conference on Earthquake Engineering in Britain, University of East Anglia, Thomas Telford Publishing, London, 1985, April, pp. 169–181.
[2] Irving, J. *Earthquake Hazard in Britain. Ibid*, pp. 261–277.
[3] Ove Arup and Partners: personal communication.
[4] Charles, J. A. et al. *An engineering guide to seismic risk to dams in the United Kingdom*. Building Research Establishment, Garston, UK, 1991, i–vi and 1–64.
[5] Principia Mechanica Ltd. *Ground Motion Specification*. Report No. ET17 for British Nuclear Fuels Ltd, 1982.
[6] International Commission on Large Dams (eds.). *Selecting seismic parameters for large dams*. Bulletin 72, ICOLD, Paris, 1989, pp. 43.
[7] Department of Environment, Transport and the Regions (DETR). *An application note to an engineering guide to seismic risk to dams in the United Kingdom*, Institution of Civil Engineers, Thomas Telford Ltd, London, 1998, 1–40.
[8] Musson, R. M. W. and Winter, P. W. Seismic hazard in the UK. Report for the DTI, by AEA Technology, Warrington, 1996.
[9] *Design criteria for concrete arch and concrete gravity dams*. US Bureau of Reclamation Engineering Monograph No. 19, US Government Printing Office, Washington DC, 1977.
[10] Ambraseys, N. N. and Menu, J. M. *Earthquake induced ground displacements*. Earthquake Engineering and Structural Dynamics, 1988, **16**, pp. 985–1006.
[11] Lafitte, R. *Probabilistic risk analysis of large dams: its value and limits*. Water Power and Dam Construction International. March 1993, Vol. 45, pp. 13–16.
[12] Gosschalk, E. M. et al. *An engineering guide to seismic risk to dams in the United Kingdom and its international relevance*. Soil Dynamics and Earthquake Engineering, 1994, **13**, pp. 163–179.
[13] Neilson, N. M., *BC Hydro's approach to dam safety*. Water Power and Dam Construction International. March 1993, Vol. 45, pp. 39–44.
[14] Hinks, J. L. and Gosschalk, E. M. *Dams and Earthquakes — a review*. Dam Engineering, 1993, February, IV (1) 9–26,.
[15] Dobry, R. and Alvarez, L. *Seismic failures in Chilean tailing dams*. Proceedings of American Society of Civil Engineers, 1967, SM6, November.
[16] Larrain, G.N. *Seismic behaviour of some Chilean earth dams*. International Commission on Large Dams, New Delhi, 1979, Q51, R23.
[17] Retamal, E. et al. *The behaviour of earth dam in Chile during the 1985 earthquake*. Proceedings of the 12th International Conference on Soil Mechanic and Foundation Engineering, Rio de Janeiro, 1989, 3, pp. 1995–2000.
[18] De Alba, P. A. et al. Analysis of dam failure in 1985 Chilean earthquake. *Journal of Geotechnical Engineering*, proceedings of American Society of Civil Engineers, 1988, December. Vol. 114, No. 12, pp. 1414–1434.
[19] Bolton Seed, H. Considerations in the earthquake resistant design of earth and rock fill dams. 19th Rankine Lecture, *Geotechnique*, 1979, 29(3), pp. 213–263.
[20] *ibid*.
[21] Resendiz, D. et al. El Infiernillo and La Villita Dams; seismic behaviour. *Journal Geotechnical Engineering Division*, Proceedings of the American Society of Civil Engineers, 1982, 108, January, pp. 109–131.
[22] Chopra, A. K. and Chakrabati, P. (1981). Earthquake engineering and structural dynamics. Vol. 9, No. 4, Jul–Aug 1981, pp. 363–383.
[23] Chopra, A. K. and Gupta, S. (1982). Hydrodynamic and foundation interaction effects in frequency response functions for concrete gravity dams.

Earthquake engineering and structural dynamics. Vol. 10, No. 1, Jan–Feb 1982, pp. 89–206.
[24] Herzog, M. A. M. Simplified analysis of concrete gravity and arch dams subject to earthquakes. Proceedings Institution of the Civil Engineers, Part 2, 1982, 73 March 189–198.
[25] Ahmadi, M. T. and Khoshrang, G. H. (1990). Dam engineering, Vol. III, No. 2, pp. 85–115.
[26] Herzog, M. A. M. Water Power and Dam Construction, August 1980, Vol. 32, pp. 28–30.
[27] Raphael, J. M. (1984). Tensile strength of concrete. ACI Journal, Vol. 81, No. 2, pp. 158–165.
[28] International Commission on Large Dams (eds.) Earthquake analysis procedures for dams: state-of-the-art. Bulletin 52. ICOLD, 1986, 148 pages.
[29] Shan, Z. T., Huang, J. D. and Wang, Z. N. Analysis of liquefaction and deformation of Douhe Earth Dam during Tangshan Earthquake. Proceedings of the Fifth International Conference on Numerical Methods in Geomechanics, Nagoya, Japan. Apr 1985, Vol. 3, pp. 1387–1392.
[30] Bolton Seed, H. Considerations in the earthquake resistant design of earth and rock fill dams. 19th Rankine Lecture, *Geotechnique*, 1979.
[31] Bolton Seed, H. *et al.* Evaluation of liquefaction potential using field performance data. *Journal Geotechnical Engineering Division*, Proceedings of the American Society of Civil Engineers, 1983, 109, No.3 March.
[32] Bolton Seed, H. *et al. Seismic design of concrete face rockfill dams.* American Society of Civil Engineers Symposium: Concrete face rockfill dams: design, construction and performance, Geotechnical Eng. Division, American Society of Civil Engineers, 1985.

Further reading

Hays, W.W. *Procedures for estimating earthquake ground motions.* Geological Survey Professional Paper 114, US Government Printing Office, 1996, Washington.

Sherard, J. *et al.* (1963). *Earth-rock dams.* Wiley, Chichester.

International Commission on Large Dams (eds). *A review of earthquake resistant design of dams.* Bulletin 27, ICOLD, Paris, 1975.

Idriss, I.M. *Evaluating seismic risk in engineering practice.* International Conference on Soil Mechanics and Foundation Engineering, San Franciso, USA, 1985, August.

Thomas Telford Ltd (eds.). *Dams and Earthquakes.* Thomas Telford Publishing, 1981. Proceedings of a conference at the Institution of Civil Engineers, London, 1980, October.

Ambraseys, N.N. *Behaviour of foundation materials during strong earthquakes.* Proceedings of 4th European Symposium on earthquake engineering, Bulgarian Academy of Sciences, London, 1972, **7**, September, pp. 11–12.

Bolton Seed, H. *et al.* Performance of earth dams during earthquakes. *Journal of Geotechnical Division*, American Society of Civil Engineers, 1978, 104, July, pp. 967–994.

Bolton Seed, H. *et al.* Analysis of Sheffield Dam failure. *Journal of Soil Mechanics and Foundations Division*, American Society of Civil Engineers, 1969, November.

Ishihara, K. *et al. Permanent earthquake deformation of embankment dams.* Dam Engineering, 1990, 1, Issue 3. Q 24, 221–232.

Swannell, N.G. *Simplified seismic safety evaluation of embankment dams, dams and reservoirs.* October, 1994.

Musson, R.M.W. *A provisional catalogue of UK earthquakes greater than $4M_L$, 1700–1900.* Report WL/90/28, Edinburgh, British Geological Survey, 1990, June.

Part II

Development practice for reservoirs

10. Water conduits for reservoirs

Purposes
Reservoirs need to be served by conduits to deliver water to wherever it is required, even if this is only from the upstream side to the downstream side of the dam. It follows that conduits are required for all the main purposes of reservoirs: water supply, irrigation and hydroelectric power development, as well as for transfer of water from the reservoir to another river or elsewhere. They are also necessary for internal purposes at the dam site or within the dam itself: for low-level outlets, spillways and river diversions during construction. Some mention was made of several of these purposes in Chapters 2, 3, 7 and 8. The size of conduit is usually decided in order to achieve the minimum present value of capital and recurrent costs of construction and maintenance combined with the present value of head losses.

Types
Fundamentally, conduits are of one or more of the following types: open, pressure conduits, pipelines or free-flowing enclosed conduits. Each is now described in turn.

Open conduits
These may be founded on cut (i.e. in excavation in undisturbed ground) or on compacted fill. They may be lined or unlined according to the velocity of the flowing water which they carry, according to the need to control losses due to seepage and leakage and according to the support provided by the surrounding ground. Due to the risk of differential movements causing serious cracking of the lining, the building of open conduits with part of the width of the conduit founded on natural ground (especially on hard ground or rock) and part on compacted fill, is to be avoided as far as possible, although at first sight the equalization of cut and fill would seem to save cost. If it cannot be avoided, special precautions need to be considered

(such as the provision of flexible but sealed joints in the lining or the use of hard materials or lean concrete for the filling).

The velocity of water in open conduits is usually as low as is found to be economic, to achieve the minimum head losses. However, enlarging a conduit to reduce the velocity incurs increasing cost and the economic optimum must be found. When water is static in a conduit, the water level throughout should be constant, whereas when flowing at maximum discharge capacity, the difference in water level between the upstream and downstream ends will be equivalent to the difference in velocity head plus friction and turbulence losses. If the upstream water level is effectively that in the reservoir, the velocity head at the upstream end will be approximately zero. Thus the sides of the conduit have to be sufficiently high and deep to accommodate the depth of flow plus freeboard at the upstream end and the depth of flow plus total head loss over the length of the conduit plus freeboard at the downstream end. The freeboard has to accommodate wave heights, including transient surge wave heights as well as any settlement of the conduit. It can be seen that these constraints add to the capital cost of a conduit and they may necessitate the introduction of drop structures or small reservoirs at intermediate point(s) along its length to reduce the depth and height required downstream.

Open conduits that are lined may, in fact, be covered if this is necessary to avoid debris falling in or to prevent unauthorized abstraction of water or, in hot countries, to reduce evaporation losses.

Pressure conduits

These normally flow full of water under pressure. They may be tunnels through soft ground or through rock, or pipes of concrete or steel or other materials or conduits of various shape (such as rectangular, egg-shaped, D-shaped, horseshoe-shaped or oval) made of reinforced concrete or steel or other materials. Through the dam itself they may be formed in concrete which is part of the dam, as outlet structures, or for a spillway. Typical layouts are illustrated in Fig. 10.1. For obvious reasons of cost, the proportion of pressure conduit which is under low pressure is as great as practicable. However, approaching a hydroelectric power station, the internal pressure must necessarily increase towards the maximum. In addition, close to and just upstream of hydroelectric generating plant, there could be high transient pressures and surge pressures due to rapid opening or closing of valves or gates, a subject which will recur presently.

Tunnels in soft ground must be lined to resist erosion and the lining must be structurally suitable to support the ground and to resist the internal pressure. Tunnels in rock need not be lined, even under high

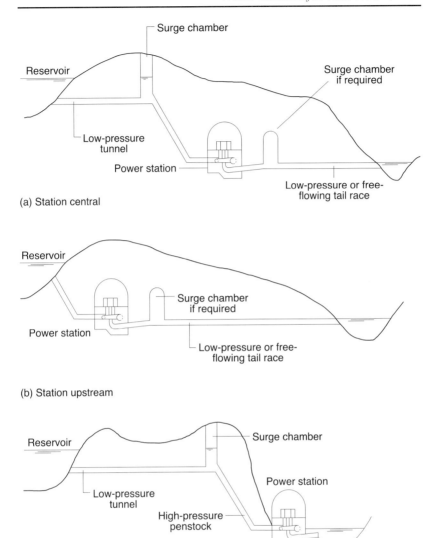

Fig. 10.1. Possible hydropower tunnel layouts

internal pressure, providing that the rock is massive, sound and of sufficient strength to resist erosion and to withstand the maximum internal pressures without fracturing. This implies that the natural stress field in the rock is sufficiently compressive before the tunnel is excavated, as will be shown later. However, if there is a risk of deterioration of the rock in the long term, tunnels, particularly for hydropower schemes, are sometimes lined to prevent future undefined falls of rock into the tunnel, which would interrupt power production, possibly for a long period, in order to allow for emptying the tunnel, effecting remedial work, and refilling the tunnel afterwards. In some cases, notably in Norway and Sweden, the quality of the natural rock is so good that even shafts and penstock tunnels under high internal pressures have been left unlined. The rock surrounding unlined tunnels can be strengthened, if necessary, by rock bolting and/or by grouting under pressure with cement.

Tunnel linings can be fabricated from the following materials.

- *Sprayed mortar/sprayed fine concrete.* This is sometimes described as 'shotcrete', which is similar to gunite. It is usually 75 mm or more thick and is often reinforced by a layer of steel mesh reinforcement, anchored in position by rock bolts. If needed to be of the highest possible strength, the shotcrete may have micro-silica (otherwise called silicon fume) additive to the cement and/or may be reinforced with steel fibres. It should be noted that the steel fibres will be embedded close to the surface, which could lead to surface deterioration if the water is acidic. The shotcrete will follow the profile of the excavation so the interior surface will be irregular, even if the shotcrete is sprayed on as evenly as practicable.
- *Precast concrete lining segments.* These are usually formed into a cylindrical shape and circumferentially compressed, using jacks. The interface between the segments and the excavation is usually grouted with cement to ensure a continuous contact and support to the ground.
- *Plain concrete.* An example of this is shown in Fig. 10.2. It is usually a minimum of 200–250 mm thick, to allow it to embed steel ribs, if these are found necessary, with a minimum of 100 mm of concrete cover over the steel on the inside face. The ample cover is advisable to allow some latitude in fixing the temporary steel (or timber) formwork required to form the internal face of the lining. This formwork is usually of considerable length, say 20 m, and collapsible to allow clearance for it to be transported along the tunnel on rails. Plain concrete is liable to crack under tension, so it is advisable to provide pressure relief holes through the lining and into the rock to equalise the pressures on either side of the lining.
- *Reinforced concrete.* The Author does not recommend the use of this type of lining because of the difficulty in placing the concrete lining round the steel reinforcement without poor compaction and the presence of voids, in a restricted width between excavated face

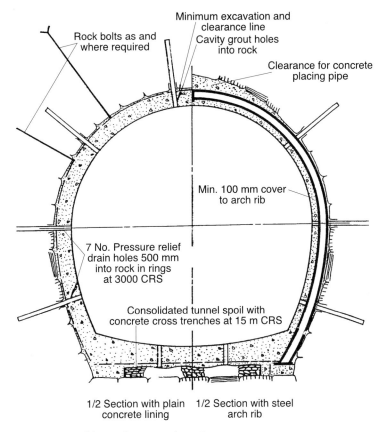

Fig. 10.2. Typical horseshoe tunnel section

and formwork. Fixing the reinforcement correctly is not an easy job and delays the operations of lining.
- *Welded steel lining*. This is usually the most reliable (but costly) form of lining to resist high internal pressures. It is usually installed in cylindrical rings, with clearance between the cylinders and the excavation into which a plain concrete outer lining can be pumped. Before placing the pumped concrete, the circumferential joints between the steel cylinders (rings) have to be welded. To avoid the extra working space for welding which would have to be excavated (and later filled with concrete), the circumferential welds can be welded entirely from the inside of the lining, as shown in Fig. 10.3, the weld being contained by an overlapping backing strip welded onto the outside of one of the two meeting rings. The finished longitudinal welds (which are likely to be fully

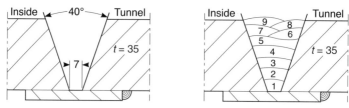

Fig. 10.3. Sketch showing welding method for steel lining from inside tunnel only (circumferential joints). t is timeless = mm.

stressed by circumferential tensile stresses when under maximum internal pressure) should be fully and reliably X-rayed before the rings are transported into the tunnel. The circumferential welds are not likely to be highly stressed and can be tested supersonically, inevitably after installation. In this way, the clearance required outside the steel lining can be kept to the minimum necessary for concreting (as shown in Figs 10.2 and 10.4).

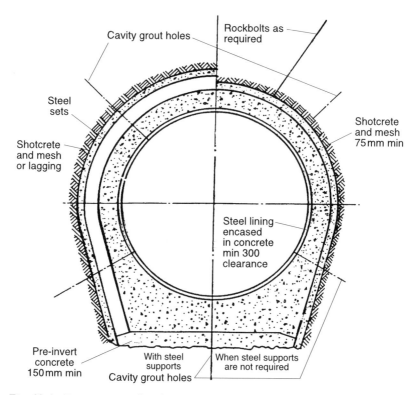

Fig. 10.4. Pressure tunnel with steel lining

Generally speaking, tunnel cross sections that are curved, resulting in mobilization of arch effects, give the best strength characteristics while minimizing stress concentrations and reducing undesirable tensions in rock and concrete. Thus for structural purposes, circular, egg, horseshoe and D shapes come out best in that order. For ease of access and clearance for transport during construction, however, except when a tunnel boring machine is to be used, they should be in the opposite order, so one chooses a shape no more curved than rock conditions are expected to demand. With the usual uncertainty as to conditions, the choice of horseshoe is often a reasonable compromise, though circular should be used if ground conditions are known to be bad. Tunnel boring machines have obvious advantages over the classical drill and blast method of excavation because they provide a structurally ideal circular shape and also a minimum of overbreak and uniform thickness of concrete lining. However, difficulty in operation in variable or bad rock conditions tends to limit their use to situations where tunnelling conditions are expected to be fairly uniform and good. It is sometimes worthwhile inviting contractors to tender for alternative tunnel shapes and methods of excavation: drill and blast or boring machine. That way one may gain the advantage of different contractors' differing experiences in different techniques and of their various available plant resources.

Pipelines

These can serve the same purposes as tunnels but they must be buried in the ground (in which case they effectively become 'cut-and-cover' tunnels) or they must be supported above ground, in which case they tend to be referred to as 'penstocks'. Reference may be made to a paper by Arthur and Walker [1]. If they are above ground, penstocks must be designed with expansion joints to allow for thermal movements and longitudinal strains. Between expansion joints and between expansion joints and anchorages, supports must be designed with sliding bearings to allow longitudinal movement while minimizing longitudinal frictional force. Anchorages, usually massive concrete blocks, must be installed at both horizontal and vertical bends, to resist unbalanced forces due to the change in direction of pressure and inertia of the flowing water. Similarly, the supports have to be designed to resist the shear forces and overturning moments imposed on them by these strains and movements in the pipeline.

The corrosion protection of the external surfaces of exposed pipelines is important and should be designed to minimize heat absorption and hence thermal movements. That of the internal surfaces is probably even more important because of the problems of access during operation of the pipeline. For steel, a system which can be recommended comprises approved shot blasting followed by application, within 6 hours, of a single coat of 'high build' coal tar epoxy paint, built up to a minimum thickness of 250 µm. Three coats, building up to give a total dry film thickness of 375/450 µm, would be better still but involve delays in application and higher costs. The application of a primer after shot blasting is to be avoided, if possible, because of the slow release of solvents from the primer, which might cause blistering of the coal tar epoxy on top.

Free-flowing enclosed conduits

These are a type intermediate between open and pressure conduits, the water surface being exposed to atmospheric pressure and flowing under gravity. The invert is usually given a gradient corresponding to the hydraulic gradient under the design flow. This type of conduit can be either in tunnel or pipeline or in a reinforced concrete conduit of box or other cross section. The advantage over open conduits is that the cross section need only be sufficiently deep to accommodate the maximum flow and surface waves and surge waves, providing that the conduit is designed to withstand the internal pressure when the flow is stopped at the downstream end and the tunnel becomes full under hydrostatic pressure equal to the level in the reservoir.

Routing

Open conduits are best adopted in fairly uniform topography with only shallow slopes, in which they can follow a fairly uniform gradient on the ground. They are most economical in ground which can be excavated with fairly steep side slopes but which is self-draining. They are not suitable in steep sidelong slopes, especially where there is a likelihood of landslides. Frequent road, rail or river crossings would add considerably to their cost. Excavation in rock should be minimized, due to its high cost. Where the terrain is unsuitable for the construction of an open conduit, it may be economic to divert the route by finding a tunnel route deeper in the hillside. Although this may be longer and of higher initial cost, the risk of having to undertake substantial remedial works later due to landslides or settlement is avoided.

The routing of tunnels is very much governed by the geological conditions in the area, the aim being to route the tunnel through massive sound rock strata of uniform quality. As far as possible, the route should avoid faults, particularly faults which would cross the alignment at an acute angle, thus resulting in a longer length of fault having to be traversed during excavation than if the intersection were right angled or at a wide angle to the alignment. Faults are liable to be associated with crushed or clayey infilling (described as gouge) or with adjoining zones with open joints or fractures in the rock, which may also bring water inflows into the tunnel. A key factor in the stability of rock around underground openings is the orientation of the joints in the rock. In general it is best to avoid a tunnel alignment nearly parallel to the planes of the main joint systems or faults or fissures — because this usually results in slabs in the roof and walls with little interlock with other slabs. Joint systems tend to be orthogonal and it is well to remember that there is at least one set of three main joint systems at right angles to one another: in looking at a face of rock only two of them are visible as joints — the third is often represented by the exposed face. The tunnel alignment should be at a significant inclination to the strike of main joints in the rock.

Particularly in tropical regions, the bedrock is sometimes weathered to a considerable depth or there may be overlying layers of weak strata. In such cases it may be beneficial (for speeding the progress and containing overall cost) to align the tunnel at a greater depth, despite the increase in internal pressure and the possibly more expensive lining which this implies. One way of achieving a deep tunnel but still subject to only atmospheric or low pressure, is to site a hydropower station at the upstream end, deep underground, close to the reservoir, as shown in Fig. 10.1(b). This maximizes the length of tail race tunnel, which is usually free flowing. Tunnelling under good tunnelling conditions may be orders of magnitude less costly than tunnelling in difficult ground. It may be noted that a tunnel, being invisible from the surface except at the portals, is helpful if there would be environmental objections to above ground routes.

In long lengths of hydropower tunnel, surge chambers or shafts are needed in order to allow satisfactory governing of the turbines when discharge in the tunnel is required to vary. This allows voltage and frequency variations in the generator outputs to be kept within allowable limits. The routing of the tunnels needs to take into account the siting of these chambers and shafts in geological and topographical conditions which are suitable. The maximum water level in a surge shaft upstream of a power station will rise above reservoir level — so the shaft needs to be in relatively high ground.

At the portals, to permit rock blasting, one looks for at least one diameter or so of good, stable rock cover over the crown. If the tunnel is long in relation to the time available to build it, it may be necessary to adjust the line to allow easy access by adit (or failing an adit, to allow access by shaft) at an intermediate point or points along the tunnel, to allow excavation to be progressed from an additional two fronts or 'faces' or even more than two. However, in good uniform rock conditions in a fairly long length of tunnel, say at least 1 km, it may be economic for the contractor to use a tunnel boring machine (TBM), especially if one that is suitable happens to be readily available. In good conditions, TBMs can make very fast progress on the excavation, several times faster than is possible by drilling and blasting. After installing and commissioning, a TBM may be expected to progress at rates of up to, say, 75 m or more per week. On three drives through basalt (compressive strength 90 MPa to 130 MPa) for the 45 km long 4·95 m diameter transfer tunnel in the Lesotho Highlands water project in South Africa, the drives were reported to have averaged 85·5, 153·9 and 136·3 m weekly with the best weekly advance of 400 m, during excavation of about three-quarters of the total length to February 1994.

Unless the tunnel is to be steel lined, since concrete linings commonly become cracked, in porous or non-cohesive ground, the depth of overburden over the tunnel must be more than the internal maximum static head of water in the tunnel, otherwise there could be an escape of water to the surface, possibly causing erosion inside the rock, or when escaping on the surface above. Similarly, if jointing in the rock could permit unacceptable leakage, the internal head of water should be less than the external head of groundwater. Unless the rock is reasonably intact, the weight of rock cover on the tunnel must exceed the internal pressure in the tunnel by a safety margin or the rock will be liable to lift. Unless the tunnel is fully lined, this calls for a depth of overburden no less than about half the internal head of water in the tunnel (a less severe condition than that for avoiding leakage to the surface just mentioned). Finally, a sensible rule sometimes adopted is that net horizontal tensions at one third of the depth of overburden above the tunnel should not be as much as to cause fissures to open in the ground This implies that the minimum principal compressive stress in the rock must be greater than the maximum principal tensile stress caused by the pressure in the tunnel. As an example, to meet this requirement in fissured rock, for a factor of safety of unity and with the head of water in the tunnel equal to the depth of overburden, if the natural horizontal stress in the rock is 1/3 of the vertical compression at tunnel level, the rock cover needs to be about 5·2 times the radius of the tunnel (a rule proposed by Charles Jaeger in 1955 [2]). When a tunnel is excavated, the stresses in the rock will be redistributed, depending on the pre-existing conditions. As shown in Fig. 10.5, tangential compression two or even three times the natural vertical compression can occur at the excavated face. If tangential compression exceeds the compressive strength of the rock, rock bursts can occur, with rock being released with explosive force.

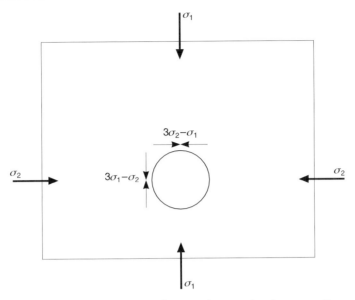

Fig. 10.5. Principal stresses around a circular opening in a two-dimensional

The guidelines just given assume a tunnel underlying fairly uniform topography. They must be strictly qualified if the tunnel underlies sloping or sidelong ground. In order to determine the minimum depth of cover L (normal to the sloping ground surface) for an unlined pressure tunnel subjected to a static head of water, H, Bergh-Christensen [3] derived the following relationship between L and H:

$$L = \frac{\gamma_w H F}{\gamma_r \cos \beta}$$

where γ_w is the unit weight of water, γ_r is the average unit weight of the overlying rock, β is the average inclination of the slope surface to the horizontal and F is the factor of safety.

For normal hydroelectric operation, Norwegian practice has been to use a factor of safety of 1·1 on the maximum static water level. This is based on the assumption that surges due to water hammer are of sufficiently short duration to have no significant influence upon the stress conditions in the rock surrounding the pressure tunnel. On the other hand, for pumped storage projects where longer duration surges may occur, a factor of safety of 1·3 is used. The Author believes that it would be reasonable to treat the maximum level reached in a surge tank or shaft under the most severe conditions of surge, as the maximum static head in this equation, with $F = 1·1$.

Permissible velocities of flow

The cross sectional area of waterway needed is governed by the rate of discharge and the velocity of flow. A lined tunnel can be smaller than an unlined tunnel, because the smooth, erosion-resistant surface of the lining allows higher permissible velocities of water without risk of damage and without unacceptably increased losses of head. For equal hydraulic capacities, Colebrook [4] has assessed the ratios of diameter of unlined to lined tunnels with equal hydraulic capacity to be about 1·3 to 1·4. The maximum permissible velocities to avoid damage to the lining are of the order:

moderately sound rock	1·5 m/s
sound rock	2·5 m/s
high-quality smooth concrete	12·0 m/s
steel	>12·0 m/s

However, head losses and imperfections in the surface often limit allowable velocities to lower values. Over long lengths of conduit, one may expect to arrive at maximum velocities of the order of 4·0 m/s in concrete and 5·0 m/s in a steel lining.

The appropriate velocity and size of tunnel in irrigation or water supply projects very often depends on the head loss possible while still retaining the operating head required to command the area to be supplied. In the case of a power project, as already indicated, tunnel sizing requires an economic study to find the marginal cost of increasing the tunnel diameter so as not to exceed the marginal present value of the energy which will be gained by this increase in waterway, due to the reduction in frictional head losses. In round terms, if the total project cost is of the order of £X and the net head is Y m, it is worth spending £X/Y to achieve every metre saving in head loss. When the tunnel is to be left unlined, it has been shown that a worthwhile reduction in head loss may be obtainable by lining the invert only (see reference [12] in Chapter 5).

The contractor needs to be told what size of tunnel to excavate — it will be expensive if it has to be enlarged later. Since, as already noted, a larger tunnel will be required if left unlined and since the rock conditions cannot be known with certainty until the excavation has advanced, for a tunnel where good rock conditions are likely, it may be worthwhile to consider ordering the excavation of the tunnel to a size which can be left unlined. Then, if rock conditions turn out to be worse than expected, a lining can be put in and head losses will be relatively low; while if the quality of rock turns out to be good enough, the tunnel can be left unlined. The penalty for this approach is that more excavation and concrete lining has been carried out if and where the tunnel has to be lined than would have been the case if the excavation had been ordered for a smaller diameter lined tunnel. It is a question of assessing the relative costs and risks of rock falls.

Tunnel lining

For rational design of the lining of a tunnel it is necessary to determine the natural stress conditions in the rock through which the tunnel will pass. Techniques for the measurement of these stresses have already been described in Chapter 4. The stresses in the rock are basically due to the weight of overlying ground and the lateral stress which this causes, constrained by the surrounding rock mass. However, past geological and tectonic processes may have built-in additional stresses or relieved existing stresses, so that the stress conditions existing are not directly related to the depth below the surface. To avoid cracks opening in a rock mass, it is necessary that the tensile stresses created in the rock by driving and pressurising the tunnel with water do not exceed the minor principal (compressive) stresses existing naturally in the rock. Since rock is a jointed material, this requirement does not rely on the rock mass having any tensile strength. Generally, over lengths where the maximum hydraulic pressure in the tunnel (including surge pressure and dynamic water hammer pressure) exceeds the minimum principal stress in the rock, the tunnel should be provided with a steel lining or other type of lining capable of carrying tension without cracking. The minor principal stress in the rock may be equated to the stress necessary to keep joints in the rock open, and can be determined directly from hydraulic fracture testing in a borehole, described in Chapter 4.

When a circular tunnel is driven in the rock in a two-directional (i.e. vertical and horizontal) stress field, the tangential stresses around the circumference of the tunnel may be illustrated as shown in Fig. 10.5. It can be seen that, if there were no horizontal compression in the rock, there would be tension in the crown of even an empty tunnel, equal and opposite to the vertical compressive stress in the rock and there would be high compressive stress in the side walls. A case history of such a problem will be described in Chapter 11. It is usually found that the average horizontal stress in rock is more than $0.5\times$ the vertical stress and at depths of less than about 140 m, the average horizontal stress is more than $1.0\times$ the vertical. However, these average values can be misleading regarding the extreme values which may be found.

What sort of lining should be selected for a tunnel? In a circular tunnel, cast in situ concrete (nowadays usually pumped into the crown within a steel shutter, without intermediate transverse joints), or grouted precast concrete segments or a steel lining encased in concrete are commonly used, a possible alternative being the 'New Austrian Tunnelling Method' or NATM. NATM is neither new nor Austrian but is an older technique which has been developed and given a certain mystique by some eminent Austrian engineers. In principle, it consists of the use of rock bolts, steel mesh and shotcrete to mobilize an arching effect in the rock around a tunnel, to prevent loosening of the rock after excavation and to provide a tough, thin, flexible lining. Early application of the shotcrete forestalls movement within the rock, reducing the potential mobilization of loading on the tunnel from the rock. For good results it needs to be constructed close to the excavated face, as soon after blasting as possible, when, in fact, conditions for good workmanship and supervision are least good and when the shotcreting operation interferes with other operations such as removal of tunnel spoil and drilling. The method has, however, been used successfully in numerous cases and the smoothing out of irregularities in the rock surface by shotcrete improves hydraulic flow conditions compared with the unlined case. A flexible lining has the advantage of allowing some deformation of the rock, thus distributing and reducing the ultimate external loading on the lining. Some fairly recent failures of underground excavations lined with the NATM, including the one under Heathrow, have raised questions about the suitability of NATM for use in soft rock, notably if the loading from the rock onto the tunnel lining continues to mobilize over a long period, rather than to stabilize.

If rock bolts form part of the design of the permanent lining, they should be grouted or fully bedded in mortar to avoid corrosion. Numerous proprietary kinds of rock bolts are available. Some have an expanding end anchor or resin anchor from which the bolt can be tensioned (to precompress the rock) before grouting. Another type (the 'Perfobolt') consists of a split perforated tube, which is filled with mortar and inserted into the hole. The bolt is driven into the mortar in the tube using a pneumatic hammer. The orientation, spacing and depth of holes must depend on the orientation and spacing of joints in the rock, but to achieve an arch effect within the rock, a bolt spacing of half the bolt length is reasonable. Bolt lengths of 2 or 3 m are quite normal. Typical empirical design recommendations (after US Corps of Engineers) have been published (Muir Wood, Cooper and Kidd [5]).

The excavation must be provided with early support where necessary to prevent rock falls before the permanent lining can be constructed (which, except in the case of segmental linings, is usually after all the excavation has been completed). Therefore, if the early support can be incorporated in the permanent lining, the external

Table 10.1. Yield strength and size of cold bent steel ribs

mm mm kg/m		mm
152 × 152 × 37	minimum radius	1525
203 × 203 × 86		3050
254 × 254 × 89		4275
305 × 305 × 118		5500

rock loads are carried without depending on the permanent lining. Steel ribs and invert struts, where used, can be regarded as reinforcement of the tunnel lining. Therefore, steel reinforcement in concrete linings can usually be avoided as already recommended, except in special situations such as transitions or, exceptionally, where required to take tensile loads due to the internal pressures. Steel ribs should be cold bent to improve the yield strength. This is a simple operation which can be carried out on site (to minimize the extra freight costs of curved ribs) but the radius to which a given size of universal column can be cold bent is limited to avoid buckling of the flanges. This limits the size of rib to be used (as shown in Table 10.1) and the spacing of the ribs may be adjusted accordingly.

If there are inflows of water into the tunnel before the lining is constructed (a severe hindrance to excavation or lining operations) grouting with cement grout is the accepted and successful means usually adopted to reduce the inflows to insignificance. However, it should first be checked whether the inflows are reducing of their own accord — they may simply result from the drainage of a limited underground aquifer, which would be drained completely by the flow reducing nearly to zero in a short time, thus saving on drilling and grouting.

Steel lining in tunnels

Competing methods have been developed for the design of steel tunnel linings [6] (see Fig. 10.4), which are adequately reported in the literature. To resist internal pressure, including surge and transient water hammer pressures (in effect shock waves due to the rapid opening or closing of a gate or valve — or to sudden rupture of the tunnel), the thickness of steel lining required will depend on the proportions of load which can be taken by the concrete placed between the steel and rock, by the rock and by the steel itself. That is, it will depend on the relative moduli of deformation of the steel, the concrete and the rock and on the external load. Grouting with cement grout under pressure should be undertaken to fill any gaps occurring between steel and concrete and between concrete and rock.

In designing against external pressure, the critical situation is likely to be with the tunnel empty and maximum water pressure in the external rock, thus maximizing external pressure, without the benefit of balancing internal hydraulic support. It is generally assumed that some form of buckling of the steel lining could take place, facilitated by separation of the lining from the concrete. The potential for buckling can be decreased and hence the allowable external loading can be increased, by stiffening the lining with external welded stiffening rings, or rings of studs. In either case, compaction of the concrete around the stiffeners becomes both important and a source of difficulty. The Amstutz theory [7], based upon buckling of the liner at a single lobe, is perhaps the most well tried analytical method.

Grouting in tunnels

Where the rock is porous or fissured and leakage is undesirable (e.g. due to risk of undue loss of water or internal erosion within the rock mass) or where the rock needs strengthening to provide better support for the tunnel, grouting with cement grout should be undertaken. This is usually carried out in stages of increasing depth and pressure: the first shallow holes at low pressure (say, 0·1 MPa close to the lining, so as not to rupture it), increasing to 0·33 MPa at depths of, say, 5 m and much higher pressure, if necessary, at greater depths. Such grouting may well be necessary where there are liable to be steep hydraulic gradients in the rock, e.g. at the locations of hydraulic gates which are used to close and empty the tunnel. Grouting pressures should be designed to avoid fracturing the rock around the tunnel perimeter, or causing uplift or leakage at the ground surface.

Final tunnel cross section

A typical final tunnel cross section is shown in Fig. 10.2. In low-pressure tunnels, the consolidated tunnel spoil may be left in the invert, being expensive to clean out and considerable in quantity because of the impossibility of aligning the drill holes for blasting closely along the invert: they have to be inclined downwards, resulting in overbreak. Drain holes in the invert are provided to prevent uplift due to external pressure when the tunnel is empty (the invert not being fully arched) and these and any other drain holes into erodible material must be adequately filtered to prevent the migration of fine material into the tunnel when the pressure in the tunnel is reduced. Concrete cross trenches are provided to prevent longitudinal percolation paths from developing in the tunnel spoil remaining in the invert. The haunches, i.e. the foundations of the walls, must be thoroughly excavated and cleaned out to rock to support the loading on the side walls and crown. If there is a risk of unacceptable leakage from the tunnel through drain holes, they can be provided with simple, non-corrodible non-return ball valves, although it cannot be expected that these will remain fully closed with the passage of time. They should protrude above the invert so that sediment does not fall into them.

References

[1] Arthur, H. G. and Walker, J. F. (1970). New design criteria for USBR penstocks. *Proceedings of the American Society of Engineers, Journal of the Power Division*, January 1970, 129–141.
[2] Jaeger, C. (1955). Present trends in the design of large pressure conduits and shafts for underground hydroelectric stations. *Proceedings of the Institution of Civil Engineers*, March 1955.
[3] Bergh-Christensen, J. (1982). Design of unlined pressure shaft at Mauranger Power Plant, Norway. *Proc. Conf. Pressure Tunnels and Shafts*, Aachen, Germany, May 1982, 531–536.
[4] Colbrook, C. F. The flow of water in unlined, lined and partly lined tunnels. *Proceedings of the Institution of Civil Engineers*, Paper No. 6281, 1958.
[5] Muir Wood, A. M., Cooper, W. H. and Kidd, B. C. Dams and their tunnels. *Water Power and Dam Construction*, Parts 1–4, February–May 1980.
[6] Jacobson, S. Buckling of circular rings and cylindrical tubes under external pressure. *Water Power*, December 1974, 400–407.
[7] Amstutz, E. Das Einbeulen von Schacht- und Stollenparazer. *Power Schweinzerische Bauzeitung*, No. 28, 1969, and *Water Power*, November 1970, 391.

11. Tunnelling problems and excavation of shafts

Low stresses in rock surround

Ideally, remedial measures should not be necessary if the design of tunnel is successful. It is hoped that enough has been said to give warning that, in tunnelling, conditions are often far from ideal and uncertainties and variations in ground conditions cannot always be foreseen. It can also happen that, during investigations and design, there are pressures on the engineer responsible to reduce costs, and solutions are adopted which would be satisfactory in the conditions predicted but which leave insufficient in hand to deal with unforeseen problems. To illustrate this, just such problems experienced on the Kotmale hydropower tunnel in Sri Lanka, which involved cracking of the lining and resulted in late changes in the design, are here described.

The Kotmale project is illustrated in Fig. 11.1. The tunnel is about 7 km long and about 6 m in diameter. When it was first put into service, some leakage developed and found its way through the rock, mainly into a construction adit and so into the drainage pit in the underground power station. The leakage gradually increased and eventually approached about 4000 l/min. At this stage it was causing concern because it was increasing and because the rate of leakage was approaching the capacity which the pumps installed in the power station were able to pump out, allowing that one pump might be out of action for repair. Drainage under gravity was not available and pumping was necessary to prevent the underground power station from becoming flooded.

During commissioning, an opportunity was taken, while the turbines were not operating, to empty and inspect the tunnel. The steeply inclined high-pressure shaft had been lined with a welded steel lining to about one third of its height, where the maximum static head would be about 200 m. Above that, the shaft was lined

186 | Reservoir engineering

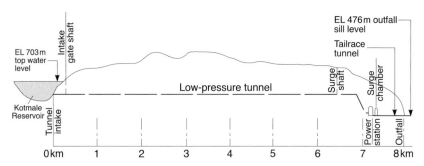

Fig. 11.1. Longitudinal section along low-pressure and tailrace tunnels of the Kotmale project

with plain concrete, with steel reinforcement over two short lengths where rock conditions found were not satisfactorily sound. The rock surrounding the concrete-lined shaft had been subjected to two-stage high-pressure cement grouting, with the aim of consolidating and waterproofing the rock, and to induce compressive support around the lining. It was found, however, that during commissioning, longitudinal cracks had occurred in the soffit and invert of the concrete lined shaft, from the top of the steel lining to the bend at the top of the shaft. This was quite unexpected, because rock conditions had been believed to be and found to be satisfactory during excavation. It was decided to make temporary repairs and to carry out measurements of the stresses existing in the rock, and then to put the tunnel back into operation, while arrangements were being made to extend the steel lining to the bend at the top of the shaft.

The temporary repairs consisted of more cement grouting round the shaft (during which it was found that the rock still accepted substantial quantities of grout, even though it had been grouted at high pressure previously) and the cracks were sealed as shown in Fig. 11.2, using an elastic sealer in a groove covered by a rubber strip. The rubber strip was held down (for example, against back pressure from within the rock) by steel plates, one on each side of the crack, which were bolted down to the concrete using resin anchored studs. This method was designed to allow substantial opening or closing of the cracks without allowing leakage.

Measurements of stresses in the rock were carried out round the tunnel, at intervals, between the steel lining and the surge shaft. The method used was that described in Chapter 4, with three strain gauge rosettes in a single borehole, generally at one depth and then at another deeper depth, after extending the borehole. The results were surprising because they indicated generally low or even tensile

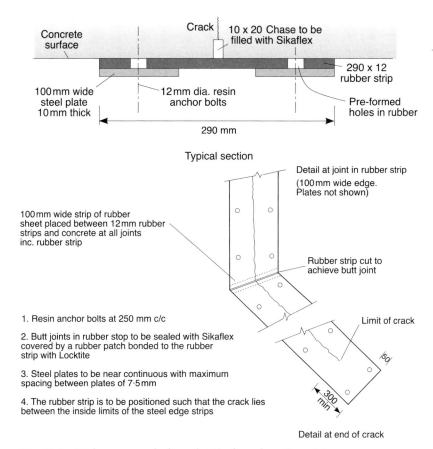

Fig. 11.2. High-pressure shaft — detail of crack sealing strip

stresses in the rock, in a horizontal direction more or less normal to the alignment of the tunnel. Even the vertical stresses measured were lower than those which would be expected from the weight or mass of the ground above. Under water pressure from the tunnel, the low principal stresses estimated would allow sub-vertical joints in the rock to open. Such joints were known to exist nearly parallel to the tunnel line. If water from the cracks in the lining escaped to these joints, if they were open, it would find its way to the power station. It has already been mentioned that measurements of stresses by strain gauge rosettes in boreholes involve inaccuracies. Later, hydraulic fracture tests indicated rather higher compressive stresses in the rock, but still much lower than originally expected. The low stresses may have resulted from erosion of neighbouring hill sides (which are steep) causing stress relief or may have been due to an anti-cline formation which exists in the rock from geological time, with possible circumferential tensions in the crown of the anti-cline.

Various studies and tests were undertaken and six alternative remedial measures were considered, to prevent or deal with possible leakage from the tunnel, as far upstream as the surge shaft and up to 2 km beyond that, where the ground cover above the tunnel became greater. The proposals were as follows.

(a) The steel lining to be extended inside the existing concrete lining up to the surge shaft.
(b) Pre-stressed concrete lining inside the existing lining (instead of the first alternative) up to the surge shaft.
(c) Drainage galleries outside the existing tunnel, combined with grouting of the rock surrounding the tunnel.
(d) A pre-stressed concrete lined bypass tunnel (which could be constructed while the existing tunnel continued in operation and would replace the existing tunnel on completion of the bypass).
(e) A free-standing steel penstock in a bypass tunnel (Fig. 11.3).
(f) Simply putting non-return valves in the drainage/pressure relief holes in the existing concrete lining, to reduce the escape of water through the pressure relief holes. This would be least expensive and quickest of all but would it be enough to reduce leakage sufficiently, considering that there were cracks in the concrete lining?

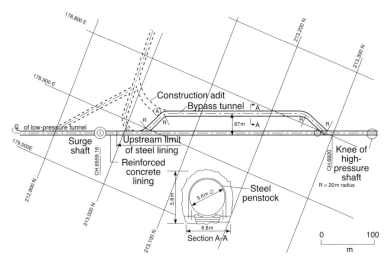

Fig. 11.3. *Steel penstock in bypass tunnel*

In each case, the rock round the tunnel upstream of the surge shaft was to be water pressure tested and grouted (if necessary), over a length of 2000 m in order to close leakage paths. In the end, as well, a tunnel for drainage and grouting was constructed parallel to the main tunnel, for a length of some 400 m upstream of the surge shaft.

The Government of Sri Lanka and the Electricity Board decided, in the end, that they wanted the most positive remedial measures possible, even despite high cost, to prevent any risk of a recurrence of the trouble and to ensure uninterrupted operation of the power station in future. It was therefore decided to adopt alternative (a) — to install a steel lining in the existing tunnel, right through to the surge shaft.

It is nearly always important and particularly so in this case, to minimize the annulus of the tunnel cross section which is taken up by the steel lining and its concrete surround. If the welding has to be carried out and X-rayed in this annulus, outside the steel lining, about 600 mm minimum clearance is required. As already noted, the longitudinal welds are the critically highly stressed welds and these can be welded outside the tunnel from both sides of the steel plate (see Fig. 10.3) and can be fully tested by X-raying. The cylindrical steel lining can then be taken into the tunnel in the longest lengths for which there is access room. Then only (some of) the less highly stressed circumferential joints need be welded inside the tunnel, from one side (the inside) of the steel plate, with a backing plate. It is sufficient to test these welds ultrasonically, since they are not structurally critical. This approach requires only a minimum clearance of 300 mm for concreting outside the steel lining, which therefore saves of the order of 600 mm in the diameter of the tunnel required, i.e. it saves 600 mm in the diameter of the excavation and 300 mm in the thickness of the concrete surrounding the steel lining. This has been dealt with at some length because on one occasion it was found extremely difficult to convince a tunnelling engineer who expected longitudinal joints to be welded from both sides inside the tunnel, as to the savings which could be made.

While these studies were in progress and decisions were being taken with the advice of numerous experts (who did not all agree on which solutions should be adopted), after the temporary remedial works described had been carried out, the tunnel was refilled and put into operation until the additional steel lining could be procured and installed. The temporary remedial works were not as effective as had been hoped, as was evident when the leakage quickly increased to about 3000 l/min, about 75% of what it had been before. Figure 11.4 indicates leakage to the power station and the hydrostatic head in the tunnel, plotted against time. A burst of leakage at about day 14 was the result of drilling a pressure relief hole near the high-pressure shaft,

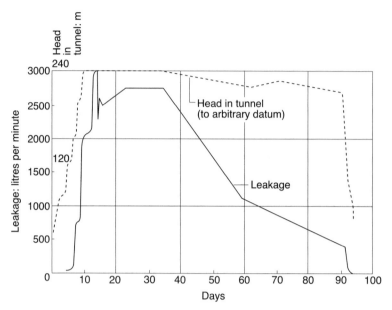

Fig. 11.4. Leakage from Kotmale tunnel to machine chamber

one of several drilled into the rock to relieve pressure in the rock around the large underground machine chamber. A drill hole evidently released a small pressurized aquifer of water trapped in the rock. Then, after a period of time, something very unexpected happened: with low in situ stresses in the rock, it had seemed likely that joints in the rock would progressively open and, also, with the erosion of leakage paths, the leakage would progressively increase. The opposite happened: the leakage progressively decreased and even under high head, reduced to the order of one third of what it had been initially under similar head. Evidently leakage paths were becoming restricted, perhaps becoming clogged by accumulation of eroded particles. It seemed possible that the leakage might be curing itself but, nevertheless, it was decided at that time to proceed with installing the steel lining. In the Author's view, the main justification for this was that the Kotmale project was designed with the possible future raising of the dam by 30 m in view. The future addition of 30 m of head might possibly trigger increased leakage once again, if insufficient preparations were done in advance.

The lessons learned from the problem were chiefly two:

- first, even where there is little reason to suspect trouble, conditions in the rock should be investigated by hydraulic fracture tests during the design stage

- second, factors of safety should not be reduced too low under financial pressure to save cost. A modest increase in the original length of steel lining in the high-pressure shaft might have resulted in restricting leakage within acceptable limits, with an overall saving in time and cost.

Two more case histories

A brief outline will be given here of the salient features of two interesting case histories in which the Author was not involved but which are described in a paper by D. C. Hitchings [1]. They give an idea of how remedial measures have to be tailor-made to an infinite variety of problems in tunnel construction. The methods largely deal with the problems caused by faults in the rock intersecting tunnels.

Victoria Hydropower Project, Sri Lanka — outfall tunnel

The excavation drive, by drill and blast methods, of the power tunnel from the downstream outfall portal, encountered heavy inflows of water which were countered by attempted grouting. However, there was an in-rush of water at the face, bringing with it rock and fault material. The inflow gradually reduced, but 'muck' continued to flow into the tunnel. Seven steel support ribs were erected to support the excavation, as shown in Fig. 11.5, and shotcrete was applied. Three exploratory boreholes, Nos 213, 214 and 215, were drilled at an incline to the line of the tunnel from behind the ribs. Drainage holes

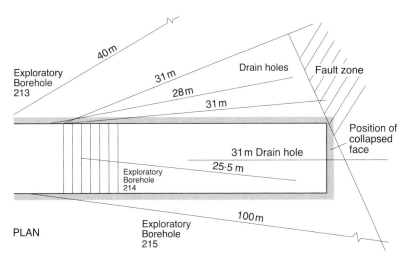

Fig. 11.5. Exploratory and drainage hole locations on the Victoria Project outfall tunnel (fault No. 1)

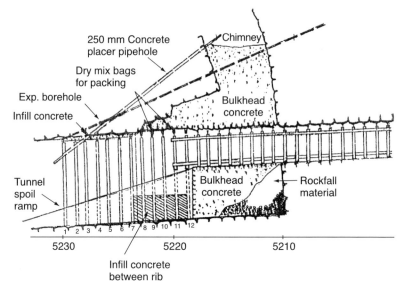

Fig. 11.6. *Victoria Project outfall tunnel — fault No. 1 support*

were drilled to reduce the water pressure in the rock. As shown in Fig. 11.6, it was discovered that an unforeseen fault had been intersected, that there was a 'chimney' in the fault and that the face of the excavation in the tunnel had collapsed.

A further five steel ribs were installed and infilling concrete was placed around the ribs. A hole was drilled to intercept the chimney, through which to pump concrete. After filling the excavation at the face and filling the chimney with concrete to the level of the placing pipe, a top heading was driven through the concrete and beyond the face (Fig. 11.6). However, having done all this, to avoid the risk of a recurrence of the trouble, a realignment of the tunnel was adopted and the intended alignment was abandoned.

The layout plan (Fig. 11.7) shows the new alignment. Probes were drilled ahead of the face to investigate conditions and to give warning of problems. The new drive had not progressed far when water inflows and fractured ground were again encountered. The excavation was supported by installing rock bolts and applying shotcrete. Drainage holes were drilled to draw water from the face, as shown in Fig. 11.8. The length of new excavation before it could be supported was reduced by shortening the length of holes drilled for blasting each round. Conditions were, however, deteriorating and the drive was progressing through very friable rock. As shown in Fig. 11.9, steel dowels were placed to support the roof of the tunnel,

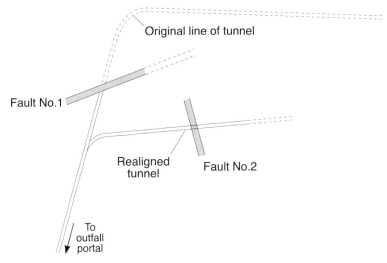

Fig. 11.7. Victoria Project — location of faults in the outfall tunnel, showing new alignment of the tunnel

in holes inclined upward and outward from the crown (a technique sometimes known as 'forepoling', which creates support for soft rock overlying a heading, by cantilevering of the dowels (sometimes of timber and called 'spiles' or 'piles')). The face was advanced 1 m under the support provided by each row of dowels.

Happily, conditions gradually improved and normal tunnelling methods were resumed. The main lesson to be learned is, perhaps, the importance of detecting in or before the design stage, the existence

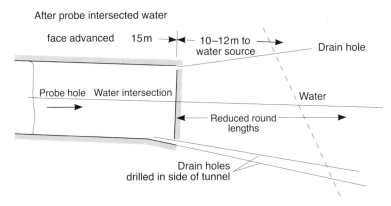

Fig. 11.8. Exploratory and drainage hole locations on the Victoria Project outfall tunnel (fault No. 2)

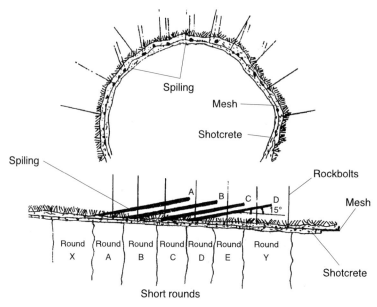

Fig. 11.9. Victoria Project outfall tunnel — fault No. 2 support

of faults which may be intersected by the tunnel, otherwise very serious problems may ensue.

Mersey Tunnel, UK

This tunnel is a 10 m diameter road tunnel from Liverpool under the river Mersey. It was to be a tunnel through rock, driven by a TBM but first a central pilot heading, 3 m in diameter, was driven by drilling and blasting. This pilot heading intersected a clay-filled fault, and the length of heading affected was supported by steel ribs. The river bed was only 4 to 5 m above the main tunnel, so a collapse of the crown would have been catastrophic.

It was decided to construct top headings to form a bridge across the fault to support the fault and the ground on either side of it. As shown in Fig. 11.10, first a central top heading was excavated and three steel beams were installed in it and then concreted in. Second, another top heading was excavated alongside the first, two more steel beams were installed in the second heading and concreted in. Third, a third top heading was constructed on the other side of the first, with two more beams concreted in. The TBM advanced beneath the beam in the crown, and the permanent precast segmental tunnel linings were successfully installed. The incident appears to illustrate a well planned, precautionary approach.

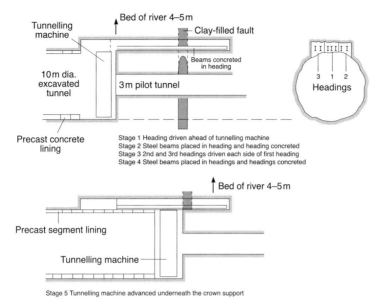

Fig. 11.10. Mersey Tunnel — method of traversing a clay-filled fault in the crown of the tunnel

Methods of excavating shafts

Usually it is quicker and more economical to remove spoil from excavated shafts from the bottom, where it can fall, than having to load it within the shaft and lift it to the top. Two methods of shaft excavation are now described.

Victoria Tunnel, Sri Lanka

The method used to excavate shafts on the Victoria (and Kotmale) Projects in Sri Lanka was by employing a raise-boring machine, as shown in Fig. 11.11: a reinforced concrete collar needs to be constructed at the surface for the full diameter of the shaft, and an access tunnel to the base of the shaft is required. Next, the raise-boring machine is positioned on the surface and a pilot hole is drilled to intersect the access tunnel. Drill rods are lowered through the pilot hole and the drilling head is installed on the rods in the access tunnel. By raising the rods and drilling, the pilot hole in this case was enlarged to 2·1 m diameter. The fallen spoil from drilling is taken away through the access tunnel. Finally, the shaft is enlarged to full size by drilling and blasting. The spoil, which drops down to the foot of the shaft, is loaded and taken away.

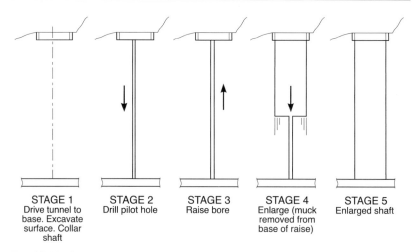

Fig. 11.11. *Victoria Project — excavation of surge shaft and gate shafts by use of a raise-boring machine*

This method is a very efficient and fast method of excavating shafts, providing that access to the base can be made available at reasonable cost. The boring equipment is, however, expensive and the cost is not likely to be justified unless substantial shaft excavation is or excavations are required. At Victoria there were three shafts: the surge shaft 21 m diameter and 104 m deep; and two 12 m diameter gate shafts, each 80 m deep. The same raise-boring machine was used on the Kotmale Project not very far away, where there was a surge shaft and a 150 m deep shaft for cables and ventilation from the underground power station. Thus the cost of mobilizing the equipment in Sri Lanka was well justified.

Foyers Project, Scotland

The method of excavating shafts on this project using an Alimak raise climber, is also described in Hitchings' paper: in this case the shaft was to be 19·5 m diameter and 83 m deep, as sketched in Fig. 11.12. A concrete collar is again needed at the surface and an adit (5 m by 3 m) was required to the base. The raise climber was installed in an enlarged excavation at the intended base of the shaft, and was used to drill and blast a 2.7 m diameter pilot shaft, working from the bottom to the top. The pilot shaft was enlarged by drilling and blasting radially, to about 8 m diameter.

As in the first method described, the rock spoil did not have to be loaded in the confines of the shaft but fell to the foot of the shaft, where it could be loaded by machine and removed. The final

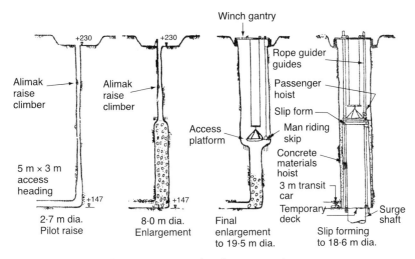

Fig. 11.12. Foyers Project — surge chamber excavation

enlargement to the full size was by drilling and blasting from the top down. This method is slower than the raise-boring machine method.

Concluding remarks on tunnelling problems

It is hoped that enough has been said to stimulate thinking on tunnel design and construction and remedial measures in 'hard' rock. 'Hard' is in fact a misleading adjective to apply to such a variable material as rock, unless it is applied carefully and literally and understood in the same way, in which case it is still a very limited and inadequate description. Tunnel design should be very intimately linked to the rock conditions through which the tunnel is to pass and it should consequently also be related to the construction methods that are suitable and expected to be used.

In this connection the overview will be ended by affirming two points mentioned above.

(a) The supports used to stabilize the rock surround during excavation are often described as temporary supports, but in fact usually remain in place in the final construction and can therefore often be an important element in the final design, if designed for permanence, which implies long-lasting resistance to corrosion.

(b) A tunnel boring machine is a very economical piece of equipment for rapidly excavating fairly long lengths of tunnel, in uniform conditions in firm to hard rock. If a TBM is used, the cross section of tunnel excavation will be circular, which is structurally a very efficient shape. Whether it will allow the quantity of excavation and lining to be minimized or not will depend on the clearances required for traffic and equipment during construction and during permanent use. It does (along with road headers), however, allow the use of a circular segmental concrete lining, with segments jacked into position, using the reaction obtainable from the tunnelling machine — a very efficient method of lining but one needing special skills and appropriate experience.

Reference

[1] Hitchings, D. C. (1985). *Journal of the Institution of Engineers, Sri Lanka*. Special Issue, Tunnelling Seminar XII No. 3, 1–64, Sri Lanka, September 1985.

12. Electro-mechanical equipment and controls

Introductory remarks
This subject is a very wide one and covers hydraulic equipment (hydraulic gates and valves and their controls), turbines and generators and their controls, transformers, switch gear and transmission lines. The subject is dealt with here from a civil engineer's point of view although much of it is very specialized and there are few electrical (E) and mechanical (M) engineers who would claim to be competent in more than one genre or sector. It is, however, valuable for both E and M and civil engineers to have an overall knowledge of the contribution that is required from those who are trained and experienced in specialized aspects. Hydraulic equipment was discussed briefly in Chapters 7 and 8, on the subject of spillways and river diversions, and it is superfluous to repeat the points already made. In view of this, a start will be made with the subject of turbines and generators.

A small number of E and M consulting engineering firms were based in the UK but two combined and have been absorbed by Mott MacDonald while the other two have been taken over by firms based in the United States: Rust and Parsons Brinckerhoff. Other firms have a good capability in E and M engineering: Balfour Beatty, through Engineering Power Development Consultants, and Sir William Halcrow and Partners Ltd also has in-house E and M capability. There are also some very small firms or individuals working on their own. Firms manufacturing hydraulic equipment and/or turbines have also become reduced to a very small number. Only Boving is able to offer equipment above the small category and Boving has become Kvaerner Boving, a member of a large European group. However, Neyrpic (French) and Toshiba (Japanese) have bases in the UK. In Europe and Japan there are many experienced manufacturers that are capable of supplying items up to the largest available, while in the UK there are a few manufacturers who can supply micro, mini and/or small turbines. The supply position for generators and electrical equipment in the UK is a little more competitive, with GEC Alsthom and NEI Peebles at the large end.

For initial planning and design, reasonably comprehensive information can be obtained from manuals such as *Civil Engineering Guidelines for Planning and Designing Hydroelectric Developments* [1]) and *Hydropower Engineering Handbook* by Gulliver and Arndt [2] as well as from earlier publications. Most manufacturers will supply site-specific proposals and advice on the design of, and supply and installation of, equipment which they themselves can offer. They should be given to understand that, if they are engaged to supply the equipment, they will be expected to guarantee the efficiency and output characteristics which they tender, to be demonstrated by commissioning tests. Nevertheless, it is advisable to engage the advice, for design, installation and commissioning, of independent E and M specialist(s) with appropriate experience, since to rely on assistance from a potential contractor is inhibiting if tenders are to be invited on a competitive basis from other firms.

Turbines

Since a turbine and its generating equipment operate as a unit, the main contractor for their supply should be responsible for both and for their combined performance, even though these may be sub-contracts for the supply, installation and commissioning of the equipment which is outside the main contractor's manufacture, to another firm. The supply should be complete with all equipment to the switch gear output terminals, including step-up transformer(s), control equipment and cabling and the output should be measured to that end point. For the measurement of efficiency, it is, of course, also necessary to provide means of measuring water inflow to the turbine(s) and net head at the turbine inlets.

For a power station which is connected to a grid, there may be sufficient stand-by capacity in the system without installing a 'spare' stand-by set in the particular station. If, on the other hand, the station is isolated or the sets are large, it may be necessary to install spare capacity equal to that of the largest unit, in order to fulfil demand when one set is out of commission or under maintenance (it can be expected that a unit would be out of service for some two weeks each year for minor repairs, but an overhaul once in, say, four years might take up to about two months). Taking into account the proportion of the total installed capacity in an isolated station which would not be productive, it is usually economic to provide about four units, including the spare; that is, one third more capacity than is required to meet maximum demand. This course of action would ensure that the sets are not 'one-off' for the manufacturer and would keep the station to an economic size.

It may be noted that the superstructure of a power station needs to incorporate a loading and erection bay, of dimensions at least equal to the clear dimensions provided for one unit, in order to provide space for dismantling and assembly of the equipment and to load it onto a trailer for removal, if required. The loading bay is usually at one end of the station with entry for a heavy trailer, under the travel of the overhead crane. The ASCE [3] guidelines recommend a length for the loading bay 7·6 m (25 ft) in excess of that of the bay for one unit, in order to allow for complete assembly of stators which cannot be shipped shop-assembled, together with space for the assembly of at least one turbine runner, as well as to allow for loading under cover. The resulting length of the loading and erection bay is expected to be about 1·5 times the width of the monolith required for one unit. It does seem that there is scope for economy in the effect of this on the overall length of the power station if, for example, it can be arranged that the loads will be lifted by the overhead crane and the trailer backed underneath for loading. It can be seen, however, that the size of the power house and its superstructure is fundamentally governed by the size and number of units to be installed. It is also governed by the type of units adopted.

The characteristics of a turbine are conventionally indicated by a quantity known as its 'specific speed'. This is awkwardly defined as 'the speed in r.p.m. at which the turbine would run at best efficiency for full guide-vane opening, under a head of one foot, its dimension having been adjusted to produce one horse power'. The specific speed, n_s, describes a specific combination of operating conditions that ensures similar flows in geometrically similar machines:

$$n_s = nP^{1/2}/H^{5/4}$$

where n is in r.p.m., P is in horsepower and H (head) is in feet. In metric units, P is either in metric horsepower or kilowatts and H is in metres. Furthermore, n_s metric $= 4.45 n_s$ British (if P is in metric horse power) or $2.83 n_s$ British (if P is in kilowatts). Hopefully this means more to E and M engineers than it does to this Author.

The alternative types of unit available and their fields of use will now be described briefly.

(a) *Francis turbines.* The Francis turbine is known as a 'reaction' type, in which the runner receives water under pressure, in an inward radial direction and discharges in an axial direction. The water enters from upstream through a spiral casing, through adjustable guide vanes

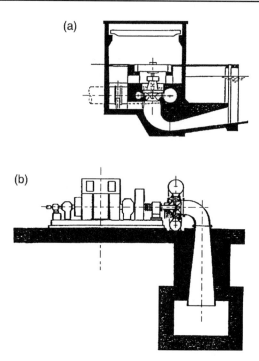

Fig. 12.1. Vertical (a) and horizontal (b) Francis turbines

and into the fixed-vane runner, where it changes direction to axial and passes out through a draft tube (Fig. 12.1). The draft tube provides a gradual enlargement of the waterway in order for the discharge to recover much of the velocity head which would otherwise be lost. The shaft for turbine and generator may be vertical (a) or, for small machines, horizontal (b). The latter can effect savings in the overall cost of the power station. The Francis turbine is suitable for heads in the range of 10–306 m (possibly even 2 to 1000 m) and power ratings about 0·1 MW to 838 MW (possibly to 1550 MW, as reported at Tennesvan Project, 225 m head, Norway).

(b) *Kaplan turbines.* The Kaplan turbine is an axial-flow reaction turbine of a propeller type with adjustable blades. As in the Francis turbine, the water enters a spiral casing, flows through adjustable guide vanes and after passing the runner blades, flows through a draft tube to the tail race. Due to the adjustable runner blades, the turbine has a high efficiency over a wide range of heads and output. It is thus suitable for low heads and relatively small reservoirs where there would be a wide range of head on the machine (Fig. 12.2). It is capable of efficiencies of 80% or more

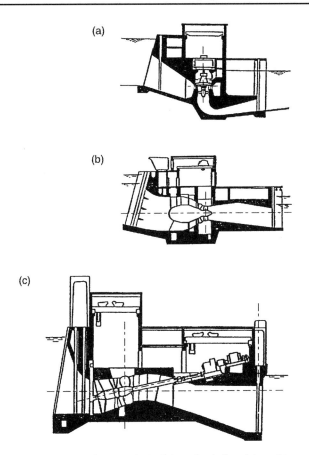

Fig. 12.2. Kaplan (a), bulb (impulse) (b) and tubular (c) turbine

when operating down to 30% of full load. It occupies a larger space than would a Francis of the same rating and is more expensive. It is generally used for heads in the range from about 5 to 75 m (possibly 2 to 530 m) and power ratings from about 0·3 to some 230 MW (possibly to 500 MW).

(c) *Propeller turbines.* These are axial-flow reaction turbines similar to the Kaplan but do not all have adjustable blades; there are a number of varieties, for example, Deriaz, bulb, tubular or straight-flow.

(d) *Deriaz turbines.* The Deriaz turbine is a mixed-flow propeller unit of the Kaplan type, originally developed as a reversible machine for pumped storage applications, but has proved itself well in the medium-head range. It provides a flat efficiency curve over a wide range of power output, similar to Kaplan propellor units but suitable for higher heads.

(e) *Bulb turbines.* In a bulb turbine the axial-flow turbine and generator are accommodated in an enclosure within the water passageway itself, with the generator being contained in a capsule. The bulb units may have fixed or movable guide vanes and runner blades. Problems to be overcome are cooling of the generator and access to the generator, unless it can be removed for maintenance (Fig. 12.2(b)). The generator must be of restricted diameter and hence low inertia, so applications are restricted to connections to electrical systems large enough to maintain electrical stability. There are believed to be cost savings in the use of bulb units compared with the Kaplan, in the range for which they are suitable, of

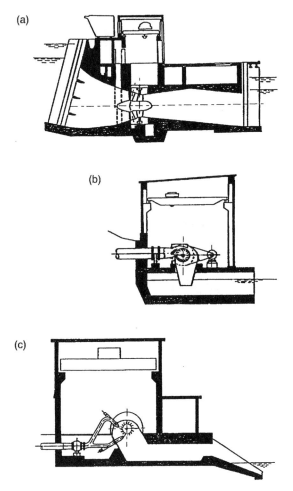

Fig. 12.3. *Straight-flow (a), cross-flow (b) and Pelton (c) turbine*

the order of 25% in the cost of the civil works for the power station and 15% in the cost of the set. There may also be improved efficiency and reduced risk of cavitation. Bulb units can be designed to operate as pumps and generators in both directions, for tidal power schemes. Bulb turbines are suitable for heads in the range of about 3·3–20 m, (possibly 2–40 m) and power ratings of more than 0·3 MW up to 54 MW (possibly even 100 MW).

(f) *Tubular turbines.* In this design (Fig. 12.2(c)), the generator is located outside the water passageway, with a long shaft drive; the generator is therefore easily accessible for maintenance. They are suitable for heads of up to about 15 m and power ratings of 0·05 MW (50 kW) and upwards.

(g) *Straight-flow and 'Straflo' turbines.* The generator in this type (Fig. 12.3(a)) is located on the periphery of the runner, the turbine and generator forming a single unit, allowing a large generator and large rotational inertia. Special seals have been developed to prevent leakage of water into the generator and into the main bearing housings (Fig. 12.4). This is a compact arrangement with

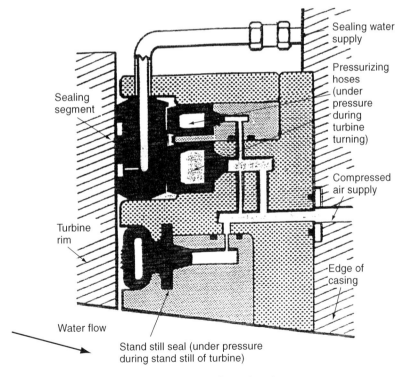

Fig. 12.4. Details of rotor rim seal (Straflo turbine)

no drive shaft, with savings likely in the extent of the civil works. Units are suitable for reversible operation as either turbine or pump. The design has been promoted for installation in large tidal power projects. Straflo (developed by Escher Wyss in 1974) units are suitable for a head range of about 2–40 m or more and power ratings in excess of 0·5 MW up to 20 MW or more.

(h) *Cross-flow turbines.* The cross-flow turbine is a radial/impulse type of low-speed turbine. They have simple blade geometry and lower construction costs but larger dimensions at low heads and high flow than the corresponding conventional turbines. The design lends itself to manufacture in conditions of intermediate technology in developing countries. Efficiency is modest but the efficiency curve is flat over a wide flow range. The type is suitable for a head range of about 2–200 m and power ratings up to about 1 MW (i.e. micro- to mini-hydro), or possibly up to 2 MW (Fig. 12.3(b)).

(i) *Pelton turbines.* Pelton turbines (Fig. 12.3(c)) have impulse wheels on which are mounted cup-shaped buckets, which have a radial partition or splitter in the centre, to divide the impinging water jet or jets, which issue from a nozzle or nozzles on the downstream end of the penstock. The wheel is encased to prevent splashing. The governing mechanism is an adjustable spear or needle and a jet deflector. Pelton turbines are suitable for high heads in the range 20–1233 m (possibly 2–2000 m) and ratings from 10 kW–423 MW (possibly to 500 MW).

(j) *Turgo turbines.* These are impulse machines actuated by a water jet, entering on one side of the runner and discharging from the other. They are suitable for heads up to about 300 m. So far they appear to have been used only in small sizes.

(k) *Pumps running in reverse.* These are commonly employed in micro- and mini-hydro installations, being generally less expensive than conventional turbines. The pumps are usually centrifugal type, which lack guide vanes. Consequently other means have to be used for starting, stopping and loading the sets, for example by adjustment of inlet valve opening.

The decision on what type of turbine to adopt and whether to accept a simple design with a very limited range of high efficiency or to accept the expense of a more complex machine with a broad efficiency curve, will depend on the expected operation of the plant with regard to variation in head and variation in the discharge required to meet load demand.

It can be seen from the above that the limits of head and capacity for different types of equipment are not hard and fast and when necessary, manufacturers should be asked for the potential capacity limits of the equipment they design and/or supply.

Generators

Two types of generator are used in hydropower installations: synchronous or asynchronous (induction). The synchronous types are normally connected to an isolated system or a grid. In the case of a grid, synchronizing equipment is required. Induction generators obtain excitation from the system to which they are connected (which should therefore be of large capacity in relation to the capacity of the induction generator, in order to ensure stable regulation) and therefore they cannot run completely isolated without the installation of separate excitation equipment. Their inertia is lower than that of comparable synchronous generators, unless provided with flywheels. Under the most suitable conditions, induction generators are more economic than the synchronous type, since they avoid the need for a separate excitation system (exciter), voltage regulator and expensive synchronizing equipment (e.g. an oil-pressure-operated speed governor on the turbine is not required). They probably require less maintenance and are less costly. Their efficiency, however, is somewhat less and there are also speed and output limitations. They are largely applied where suitable in the range 0·5 MW upwards.

Except in the case of units which are large in relation to the size of the system to which they are connected, governors are an important requirement in order to achieve constant frequency, by controlling the speed of the turbines to match changes in load in a sufficiently short period of time. They also facilitate starting and stopping and the sharing of load between units. Their operation is normally based on an actuator to sense speed changes and a servomotor to produce the forces required to adjust turbine guide vanes, runner blades, etc. An alternative to the conventional mechanical or electronic governor increasingly employed, is the use of a microcomputer to control the output of the unit. The microcomputer can also take on other tasks to improve operational control.

Transformers

Large units are usually directly connected to their own step-up transformers. To minimize the length of low-voltage/high-current cabling required, transformers may be located as close to the power station as possible — over the draft tubes or sometimes on the roof or, especially in underground stations, in a chamber adjoining the machine chamber. Sometimes two-stage transforming is adopted, low-voltage step-up in or near the station and high voltage in or near the switchyard at some distance. It is a matter of economics, as affected by topography and a matter of fire risk. Unless there is sufficient clearance and isolation for overhead cables, high-voltage cables are usually in an oil-filled casing and require appropriate design to achieve the necessary insulation. However, conductor size, temperature rise and energy losses are higher for low-voltage cables.

Transformers, stators and inlet valves are usually the largest and heaviest components to be transported and installed for a hydropower scheme, and are therefore items for which data on weights and dimensions are required from manufacturers. If road and bridge capacities and clearances are likely to affect the feasibility of transporting the largest items, the feasibility and additional cost of breaking the items down into integral parts, which can be reassembled on site, should be discussed. Even so, prior to transport, trial shop assembly at least should be required, to ensure satisfactory accuracy of fit for erection on site.

Main travelling cranes

In the power station, an overhead travelling crane is required, primarily for handling the turbine and generator. The crane is usually mounted, accurately aligned and positioned on rails, mounted on corbels of crane columns built into the walls. The lifting height required depends not only on the unit type and sizes but on provisions

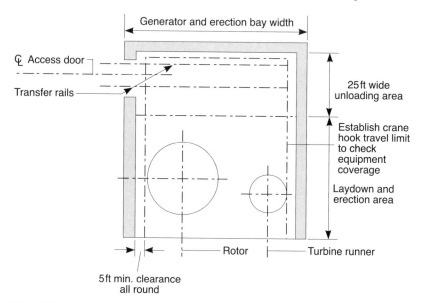

Fig. 12.5. Minimum unloading and erection space

built into the design for installation and removal. The setting of the turbine runner is constrained to provide a minimum submergence below minimum tailwater level, to achieve adequate operating efficiency at minimum head and to avoid cavitation of the runner. Depending on the arrangement and design of the bearings for the turbine, it may be possible to arrange for the main shaft to be divided, so that the turbine can be withdrawn at low level as a separate unit. If so, the route for withdrawal to the loading bay needs to be carefully planned, with all the clearances necessary. In this case, clearance has to be provided for the crane to lift the turbine runner and shaft clear of the unloading area

Span: ft	A	B	C	D (in)
200 T Crane				
60, 70, 80	13–6	12–4	8–0	13
90	13–9	12–6	8–6	13
300 T Crane				
70	16–6	15–0	8–6	15
80	16–9	15–3	8–6	15
90	17–3	15–6	9–0	15
400 T Crane				
70	17–3	16–9	11–6	17
80	17–6	17–0	11–6	17
90	18–0	17–0	11–6	17
500 T Crane				
70	18–9	18–0	19–3	20
80	19–0	18–0	19–6	20
90	19–3	18–6	19–6	20
600 T Crane				
70	20–0	19–3	20–6	23
80	20–3	19–6	20–9	23
90	20–6	19–6	21–0	23
700 T Crane				
70	21–6	20–6	20–6	26
80	21–9	20–9	20–6	26
90	21–9	21–0	21–0	26
800 T Crane				
70	22–0	21–0	21–0	27
80	22–3	21–3	21–6	27
90	22–6	21–6	22–0	27

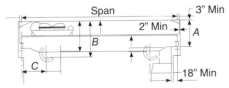

Fig. 12.6. Crane clearance data. ASCE/EPRI guides 1989. Note: D *in table not shown*

(Figs 12.1–12.3, 12.5, Table 12.1). Clearance is required below crane rails for the lifting tackle at its maximum height, and above this, for the crane gantry itself, for the control carriage, if provided, and the roof (Fig. 12.6). Of course, for small units, lifting provisions may be simplified. For large stations, crane spans of 18·25–27·5 m and capacities of 200–800 tonnes or more may be contemplated [1].

Transmission systems

Design of new and optimization of existing transmission systems is a subject requiring specialist knowledge and experience. Losses in transmission and distribution in an extensive system should not exceed some 10% of the energy input but, particularly in old, poorly maintained systems, losses can rise to several times this percentage, so there can be scope for achieving very worthwhile energy savings by reducing losses. The information shown in Tables 12.1–12.3 is for general guidance only. Standard transmission voltages are shown in Table 12.1 [3].

British Standards 77 and 3026 (1958) [4] on voltages for high-voltage transmission systems (a.c. and d.c. respectively) have been withdrawn and do not seem to have been replaced, at the time of writing, presumably pending European standardization.

"The Practical Power Limit" (P_p) is the maximum power which can be stably transmitted over an unfaulted circuit, allowing for sudden load variations. It is purely a function of the circuit considered:

$$P_p = 0.7 E_2 (Z - R)/Z_2 \quad \text{(in MW)} \ [3]$$

where Z and R are respectively the impedance and the resistance of the line in ohms and E is the line voltage in kilovolts. The Practical Power Limit is thus taken as 0·7 of the absolute transmissible power that can be stably transmitted over an unfaulted circuit, on the occasion of a switching operation or of a fault and subsequent clearing thereof on another circuit.

Transmission voltages are commonly as shown in Table 12.2.

Table 12.1. Standard transmission voltages

International: kV	British: kV
110	110
130	132
150	165
225	220
275	264
380	

Table 12.2. Common transmission voltages — advantages for distance

kV	MW	Distances
415 V	for capacities below 0·2 MW	(micro)
11 kV (or 13·8 kV)	for capacities 0·2–1·0 MW	(mini)
33 kV	for 1·0–10 MW	(small)
66 kV	for 25 MW	below 50 miles
132 kV (or 110 kV)	for 25 MW	above 50 miles
132 kV	for 50 MW	10–50 miles
220 kV	for 100 MW	up to 250 miles
220 kV	for 200 MW	50–250 miles
220 kV	for 400 MW	70–125 miles
275 kV	for 400 MW	125–300 miles
275 kV	for 800 MW	70–170 miles
400 kV	for 800 MW	170–400 miles

Direct current transmission may have advantages for distances exceeding some 250–350 miles, justifying the high cost of conversion to a.c. at each end. It can be seen that 66 kV is stated to be suitable for transmission of 25 MW below 50 miles, but 66 kV seems to be going out of use in national systems. Thus the economic voltages, depending on distance of transmission, may be as shown in Table 12.3 (132 kV can be wooden pole for single circuit transmission, or is likely to be steel tower for double-circuit transmission).

For utmost reliability and security of supplies, the system should have two single-circuit cable transmissions (ideally on separate transmission poles or towers). This is not usually economic but a transmission ring achieves much the same effect, if power can be transmitted in either direction, thus capable of serving any demand, even if there is a fault or break in the ring. Twin-core cables are susceptible to many of the possible failures which could affect single-core cable, and simultaneous faults on double circuits are possible and must be considered. For a given line voltage and conductor size, however, two separate single-circuit lines generally cost more than one double-circuit line, the ratio between the two costs being about 1·2 for 132 kV reducing to about 1·05 for 400 kV.

Table 12.3. Economic voltages depending on distances of transmission

MW	kV
25–50	66 or 132
100–200	132, 220 or 275
200–400	220 or 275
400–800	220, 275 or 400
over 800	400

Hydraulic valves and gates

This subject has already been discussed briefly in Chapters 7 and 8. For coherence, some of the points are recapitulated here. However, these guidelines do not attempt to explain details of the design of hydraulic gates, which is an extensive and complex subject, best left to experienced manufacturers and specialists in the subject. A very good introduction to the details can be obtained from the paper 'Hydraulic Gates' presented by Lewin [5]. He has more recently written *Hydraulic Gates and Valves in Free Surface Flow and Submerged Outlets* [6]. Professor Jack Lewin lectures at City University, London, and the Author has learned much of what he knows about hydraulic equipment on several projects on which they have worked together. A useful summary, Hydraulic gates: the state of the art, was given in a paper by Erbiste [7] in *Water Power and Dam Construction*, April 1981, including tables of the then largest gates of various types, from which some of the following information relating to large gates has been drawn. The journal *International Water Power and Dam Construction* publishes annually a yearbook with tables of periodic data on the civil, electrical and mechanical components of large hydropower plants, including recent contracts with sizes and dimensions, etc. [8]. The technology is, of course, developing all the time but these publications incorporate a history of comprehensive information — and principles and parameters have not changed fundamentally. Dimensions quoted should not, however, be taken as limiting without confirmation when required. They are intended only for guidance.

Gates and valves can be categorized both by their functions and by their type. Some gates of a given type can serve more than one function. The gates should always have been sold to serve the limiting operating conditions with regard to variations of head and flow, opening and closing and energy dispersal on the project for which they are required.

The choice of type of gate to serve a particular function or functions under given operating conditions, is usually a question of minimizing costs, providing that the choice is technically sound. The cost of new equipment can usually be minimized by economies in design, so it is prudent to consider whether the operating requirements could ever become more severe than first expected, since the cost of equipment to fulfil the most severe conditions might not be nearly as great as the cost would be to modify the equipment later, if found necessary. There are five main functions; each is now described in turn.

- *Function A. Guard gates or valves or stop logs* (to allow emptying of the waterway, for maintenance or repair 'in-the-dry' of gates or valves or equipment installed on their downstream side). It is usually economic to use equipment for this purpose which is not designed to

operate (to open or to close) against full unbalanced head, in which case, a relatively small bypass conduit is required, which can be outside or through the gate leaf, with a guard valve and control valve near its upstream end. The bypass valves will be submerged but must be operated in-the-dry to refill the conduit, so remote control may be necessary. Reliable operation is essential because if the bypass valves will not operate, it may be impossible to fill the conduit in order to open the guard gate, in which case the waterway would remain closed and inoperable. With suitable design, gates or valves can be 'cracked open' against unbalanced pressure, to allow a small, high-velocity flow of water past the lower lip of the gate, to fill the conduit downstream, so avoiding provision of a separate bypass. This requires design to avoid vibration of the gate and probably stainless steel sill and side plates will be required in the conduit, to avoid erosion. This is usually economic and reliable in the case of vertical lift gates, but if the gates are not designed to operate at greater partial openings, wider opening may cause vibration and must be avoided.
- *Function B. Emergency gates or valves* to close very rapidly against full head, in the event of a failure downstream. They may not necessarily need to be capable of opening against full head, which requires a considerable lift to overcome self-weight, friction and down drag on the gate, whereas closing can sometimes be accomplished under gravity loading, without power assistance, and may indeed need braking. However, to open under balanced pressure, as in Case A, a bypass or a design for crack opening will probably be required.
- *Function C. Control gates or valves* to regulate the full range of flow required. These must be capable of operating at partial openings without vibration.
- *Function D. Gates or valves for flood discharge* to pass controlled discharges up to a maximum rate during floods, as far as possible unaffected by sediment, flotsam or submerged debris carried by the water. This category includes spillway gates at the crest of the dam and low level valves or gates to control outlets for flood discharge.
- *Function E. Energy dispersing valves* to dissipate the high-velocity energy of their discharge without causing damaging erosion or other damage. Generally the energy is dispersed by the creation of a diffused jet, which becomes aerated.

The types of large gates used to fulfil functions A to D are generally as shown in Table 12.4.

Table 12.4. Hydraulic gates used to fulfil functions A to D

Type of equipment	Functions	Maximum span: m	Maximum height: m	Maximum head: m
High-pressure gates				
Fixed-wheel gates	A to D	12	19·1	96 (exceptional 150)
Tainter gates	A to D	16	13·2	110 (exceptional 136)
Slide gates	A, B & D	7·1	7 (exceptional 11·6)	189 (exceptional 286)
Caterpillar gates	A to D	11·6	15·75	105
Underflow spillway gates				Hydrostatic load: MN
Tainter gates	A to D	28·4 (exceptional 43)	21·3	46
Fixed-wheel gates	A to D	30 (exceptional 55)	20·2	25
Slide gates	A, B & D	12·8 (exceptional 20)	10·5	5·8
Overflow spillway gates				
Flap gates	C & D	56 (exceptional 100)	6·9	6·9
Drum gates	C & D	41 (exceptional 91)	9·0	14·8 (exceptional 32·9)
Sector gates	C & D	54	8	9 (exceptional 12·3)
Bear trap gates	C & D	34	5·5	2·4 (exceptional 5·1)
Overflow & underflow spillway gates				
Tainter gates with flap	C & D	26 (exceptional 45)	17	32·2
Double-leaf fixed-wheel	C & D	30	16	30·7
Fixed-wheel with flap	C & D	41·6 (exceptional 50)	11·75	12·4

Note: The double-leaf fixed-wheel type of overflow and underflow spillway gate has an advantage over other vertical lift crest gates, in that it does not require such a high lift as would a single-leaf gate and thus can result in a more economical height of piers and bridge, from which to hoist the gates and to store them above water level. It requires double guide grooves. Underflow radial gates can also be provided with an overflow flap, hinged at their top level, which gives more sensitive control of low flows and assists in discharge of flotsam.

In the design of hydraulic gates, very careful attention must be given to:

- ice loading
- loading from floating trees or logs (if either are possible)
- provision of long-lasting protection against corrosion (because of difficulties in obtaining frequent access for painting)
- avoidance of features or profiles which could lead to vibration (guidelines on this are given below)
- the efficiency of seals against leakage (which can cause damage as well as loss of water)
- accuracy in installing guides, grooves and bearings
- the provision of limit switches (and back-up for them in case of their failure) in order to prevent over-travel of gate or operating gear
- providing sufficient air entry to the discharge of high head gates (or valves) to prevent the formation of air voids and explosive sub-atmospheric pressures (particularly likely if a conduit is filled too quickly downstream, or with too little air entrainment, although release of air 'balloons' in the crown (by venting) may also be necessary)
- control systems standby facilities.

Reference has been made to valves in the case of low-level flood discharge outlets in Chapter 7. There follow some comments on different types of valves along with classification of their functions, A to E as used for gates earlier in this chapter.

(a) *Butterfly valves, functions A and B*. Butterfly valves are suitable for heads up to 230 m or more. They are made in sizes up to more than 3·7 m diameter. Discharge coefficients are normally in the range 0·7 to 0·8. These are probably the least costly type of valve for operation under high head but are not suitable for regulating use at partial openings. They consist of a disc centrally hinged on a diameter, with closure actuated under gravity by a counterweight. The disc can be formed in the shape of an aerofoil, with an internal structural framework, to reduce weight and to improve discharge coefficient. If required, a bypass facility needs to be separate. An air-entry valve must be provided in the soffit of the conduit downstream, to avoid sub-atmospheric pressures on rapid closure.

(b) *Howell Bunger valves (Fig. 12.7), functions C, D and E*. Energy dispersal is by discharge through an annulus to form a conical aerated jet. They are vulnerable to obstruction by debris unless well screened. A conical hood may be required to contain the spray, and if so, care should be taken to avoid risk of vibration of the hood, for example by anchoring it to a concrete support.

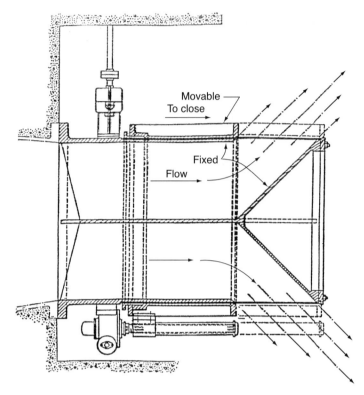

Fig. 12.7. Howell Bunger valve

(c) *Hollow-jet valves, functions A, C and D.* Maximum head approximately 300 m. They are made in sizes up to 2·75 m diameter or more. Approximate discharge coefficient of 0·7. They are also vulnerable to obstruction by debris unless well screened.

(d) *Vertical lift valves, functions A–D.* There are various types, similar in principle to vertical lift gates, the leaf being withdrawn into an enclosed hood.

(e) *Spherical valves, function C.* Maximum head about 550 m or even more. They are made in sizes up to more than 3·0 m diameter. The discharge coefficient is 1·0. These are expensive, but smooth in operation and provide a completely unobstructed waterway when fully open. They are very suitable as turbine inlet valves.

Protection of gates and valves against corrosion

Due to the difficulties of frequent and regular access, protection of hydraulic equipment against corrosion is very important and starts with the basic design. The corrosive characteristics of the water to be discharged should be determined by sampling and laboratory analyses. Possible long-term changes in the characteristics should be estimated. (Conditions after filling a reservoir are commonly likely to be acidic, due to decomposing vegetation).

Encrustation with, e.g., barnacles or other phenomena that could obstruct gate guides and so on is also a possible hazard to be considered. Cast iron and mild steel are subject to nodule growth, particularly in deeply submerged, sluggish, peat-laden water. The thickness of nodules may rise to the order of 20 mm. While soft at first, they become hard if undisturbed. Paint or galvanizing does not prevent nodule growth, but nodules do not adhere to gunmetal or stainless steel (or to any metal exposed to high-velocity turbulent flow).

The use of stainless steel and/or various alloys for critical components that are underwater and permanently lubricated bearings is recommended. Where paint treatment is acceptable, high-build coal tar epoxy paint to a thickness of minimum 250 µm in a single coat on shot blasted steel is recommended.

Avoidance of vibration

It is much better to avoid the risk of vibration by sound design than to have to deal with vibration problems after a gate is put into operation. In the worst cases, vibration can lead to resonance and instability. The necessary precautionary measures include the following.

- Providing the gate with sufficient structural strength and rigidity to withstand hydrodynamic loads, including loads due to turbulence, due to vortex shedding and to downpull.
- Designing the lip on the bottom edge of undershot gates with an as sharp and clean as possible flow break-off point.
- Keeping structural members in the gate as high and as narrow as possible.
- Detailing solid rectangular rubber seals at the bottom edge, for sealing against the sill, which should be of stainless steel, well anchored into reinforced concrete. The seals must be firmly clamped to the gate and should project not more than 6–10 mm below the skin plate. Hollow or even solid bulb seals should not be used in this location.
- The design of sealing arrangements should provide, wherever possible, for pressure on the seal to be assisted by the hydrostatic pressure of the water upstream.
- Avoiding the use of long suspension cables, unless satisfied that elastic extension of the cables due to hydraulic downpull forces on the gate will not lead to vibration.

- Avoiding excessive clearance between the gate, and the guide channels on each side.
- Careful design of the profile of overflow surfaces, to intrude slightly into the lower nappe of the maximum overflowing jet, in order to avoid sub-atmospheric pressures. The provision of spoilers to divide and aerate the jet at intervals and/or ventilation of the nappe can dispose of this problem.
- Arranging the geometry of radial gates, so that the line of the resultant thrust on the gate passes marginally below the centre line of the trunnion bearings (hinges), thus providing a positive closing moment on the gate rather than an opening moment, which can lead to instability and even to break-up of the gate.
- The sharp corners of guide slots should be faired, and preferably stop log or other grooves which are rarely used, should be blanked off with inserts.

Professor Lewin has remarked that, in practice, gate design to avoid vibration has to be based on study of the literature, practical experience and common sense.

References

[1] American Society for Civil Engineers (eds.) (1989). Civil engineering guidelines for planning and designing hydroelectric developments. Energy Division, American Society of Engineers, New York. 5 volumes ASCE/EPRI Guides 1989. Some chapters not included in the 1989 edition.
[2] Gulliver, J. S. and Arndt, R. E. A. (eds.) Hydraulic conveyance design. Hydropower Engineering Handbook. McGraw-Hill Inc., USA, 1991, 1st edition, chpt 5, 5.35–5.41.
[3] American Society for Civil Engineers (eds.) (1989). Civil engineering guidelines for planning and designing hydroelectric developments. Energy Division, American Society of Engineers, New York. 5 volumes ASCE/EPRI Guides 1989. Some chapters not included in the 1989 edition.
[4] BS 77 and BS 3026 (1958).
[5] Lewin, J. Hydraulic gates. Institution of Water Engineers and Scientists, River Engineering Section, 1978–1979.
[6] Lewin, J. (1995). Hydraulic gates and valves in free surface flow and submerged outlets. Thomas Telford, London.
[7] Erbiste, P.C., Hydraulic gates: the state-of-the-art. *The International Journal of Water Power and Dam Construction*, April, 1981.
[8] Erbiste, P.C., Hydraulic gates: the state-of-the-art. *The International Journal of Water Power and Dam Construction*, April, 1981. Vol. 33, pp. 43–48.

13. Environmental considerations

Introduction

Environmental considerations have, throughout history, strongly influenced the creative side of the approach of engineers to planning and design. Over the last 50 years or more this has been illustrated, for example, by the planning and design of hydropower projects in Scotland, which, it can be argued, have enhanced already spectacular scenery, made it more accessible for human enjoyment, and provided power and energy for people in remote areas at satisfactory cost and without any lasting seriously adverse impacts. Indeed, had the design of these projects not been environmentally pleasing, public and Parliamentary opposition would have prevented their development. Figures 13.1 and 13.2 illustrate Monar Dam, a 40 m high arch dam in Invernessshire, Scotland and Clywedog buttress dam, 72 m high, on a tributary of the River Severn in Wales. It is well known that increasing awareness of the potential hazards which are being and can be created for both the present and future generations, even beyond the limits of foreseeable time, has made people conscious of the efforts needed to attenuate existing hazards, if possible, and to avoid creating hazards for the future. Improved standards of health, safety and care for the environment have increasingly become the subject of government regulation worldwide. Nevertheless, economic and commercial pressures have all too often resulted in adverse effects, and serious technical and administrative mistakes have also been made. Some of the most obvious of these have been cases where rapid rates of sedimentation in reservoirs have shown that large human and financial investments were misused, with the waste of valuable land resources. Figure 13.3 illustrates the problem, with examples from Africa, Asia and America of sedimentation rates in cubic metres per square kilometre per year plotted against catchment areas. The rates plotted show average soil losses generally above 1 mm per year, while the histogram shows that, in a substantial majority of cases, the capacities of the reservoirs were reduced by

220 | Reservoir engineering

Fig. 13.1. Monar arch dam, Strathfarrar Project, Scotland

50% in less than 60 years. Other failures have been in the salination of large areas of irrigated land, due to evaporation from raised water tables or during designed supersaturation of the land. While the benefits from reservoir schemes in providing for human needs for water, food, energy and relief from flooding, have been beyond measure, indeed irreplaceable, in many cases insufficient provisions have been made for the social and economic needs of persons displaced from land inundated by reservoirs and for the welfare of such persons.

Environmental considerations | 221

Fig. 13.2. Clywedog buttress dam, Wales

The following is a summary by Dr W. Pircher, formerly President of the International Commission on Large Dams (ICOLD), of the criticisms of large dams by various environmental groups. Incidentally, ICOLD has defined large dams as being those over 15 m in height.

Most irrigation systems ruin more cultivable land than they create, and are often responsible for the spread of waterborne diseases. To generate cheap hydropower for mostly superfluous industries, vast ecosystems are irrevocably destroyed, and the economies of ill-advised Third World countries ruined, merely to fill the coffers

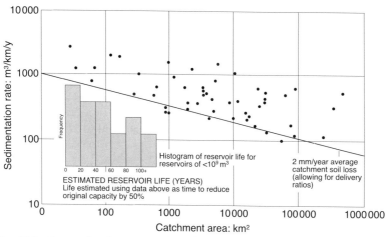

Fig. 13.3. Rates of sedimentation (catchments in Africa, Asia, North and South America). Note: see figure 4.2 — Siltation index for a more detailed continent based breakdown.

of the multinationals and the pockets of an urban elite. Flood control is no more than a myth, but reservoir induced seismic activity is a fact. Resettled populations nearly always lose their cultural and social roots, which further exacerbates their poverty. All too often, the temptations of the huge sums of money involved, create a veritable quagmire of political corruption and speculation. If all environmental impacts were properly evaluated in terms of their long-term costs, hardly any project involving the construction of a large dam would ever get further than a feasibility study. And in any case, all reservoirs — and usually much sooner than predicted — go the way of all dams and are lost to sedimentation.

Actual and potential violations of the environment by reservoir projects have led to the build-up of antagonism towards and suspicion of reservoir projects by those seriously concerned about risks to flora and fauna, risks to beautiful and treasured scenery, health and social welfare, etc. Others have joined the antagonists with less worthy motives, usually political.

It is, however, very pertinent to note that reviews of the performance of one of the large dam projects most widely criticised in recent times, the Aswan High Dam in Egypt, completed in 1968, have acknowledged it to be a successful project. Gasser and El-Gamal [1] have reported that it has made a tremendous contribution to the economic and social development of Egypt. It has protected Egypt from high floods during the seventies and saved Egypt from devastating droughts in the eighties. The dam has affected the regime of the Nile, resulting in a lowering of both water and bed levels. As a result, Esna Barrage has been replaced, a new navigation lock at Naga Hammadi Barrage has been built and many navigation bottlenecks at bridges crossing the river and along the river course are being solved [2].

The engineer's view is that to progress by change and development is part of human nature and is a central purpose of life. The avoidance of change and the doctrinaire preservation of nature exactly as it happens to be at the time in question are quite contrary to the professional training and purposes of engineers, who continually seek to improve on what has already been achieved. The achievement of aesthetic beauty, functional as it may be, the enhancement of scenic beauty and the improvement of quality of water available for man, have been and continue to be fundamental objectives of engineers. The assessment of the environmental implications of their work should always be subject to impact assessments of a satisfactory standard by suitably qualified scientists and specialists. Measures must be taken to mitigate unacceptable effects sufficiently, and the cost of such measures must be accounted for against the benefits of the project.

One particular consideration can be singled out here because it has perhaps not been sufficiently considered elsewhere: it is well established that the filling of reservoirs can, in some circumstances, induce earthquakes. Whereas dams in the area of risk should be designed to withstand such earthquakes without disastrous failure, the earthquakes might trigger landslides or cause other damage resulting in loss of life. The point to be made here is that the greatest magnitude of event ever caused by a reservoir-induced earthquake, has been about 6·5 (Richter). It is in fact credible that earthquakes of such magnitude could occur anywhere, even in regions regarded as seismically inactive, without the influence of any reservoir. Well-known examples have been at Tennant Creek, Australia (magnitudes 6·3, 6·4 and 6·7 in 1988) and at Newcastle, Australia (magnitude 5·5). Such occurrences in inactive areas are at present unforeseeable and of extremely low probability. An important conclusion to be drawn is that the probability of damage due to reservoir-induced earthquakes may be different, but the potential damage is not greater than what might occur naturally in the absence of any reservoir. The inference is that in either case, the environment should be made safe against loss of life from such events or that emergency action plans should be established to preserve life, should one occur.

Overview of the arguments for more large reservoir projects

An essential but often questioned objective of reservoir development is 'sustainability'. The World Commission on the Environment and Development in the 1987 Brundtland Report, *Our Common Future*, stated that humanity has the ability to make development sustainable 'to ensure that it meets the needs of the present without compromising the ability of future generations to meet their own needs'. This definition is both pragmatic and to the point and deserves to be widely adopted.

It can be seen [3–5] that the Author has for many years expounded the case for dams and reservoirs, developing the thinking summarized in this chapter. He has therefore been particularly conscious and dismayed by the misuse and abuse of development practice which have been experienced in some instances.

The creation of large reservoirs has been a practice and policy of man for thousands of years. As populations have increased, the demands for water, food and energy, have correspondingly increased, as well as the need for mitigation of flooding. The objective of conserving and utilizing natural water resources, which are so necessary and beneficial, is one which is inherent in human nature. In addition, as populations become more developed and civilized, their standards become higher and the use of water per capita increases and the demand for water of higher quality grows. Similarly, demand for quantity and quality of food and energy increases. In principle, small reservoirs cannot provide the regulation and efficient use of water resources that can be obtained from large ones. To achieve the same output as one large reservoir, a complex of smaller reservoirs would be required, covering an aggregated area much greater than that of the equivalent large reservoir and needing much higher runoff. With regard to flood control, small reservoirs have no effect in attenuating extreme floods and, therefore, increasing the number of small reservoirs in series in a valley does not give relief from large floods. Important indirect benefits from reservoirs can include the potential for navigation, for fisheries and for amenity use.

The benefits of reservoirs in avoiding droughts, famines, catastrophic flooding and disasters from power failures, receive little notice or quantification because, once a reservoir has been built and is performing its function, there is no necessity to go to the trouble and expense of ascertaining what would have happened had the reservoir not been built. Back-analysis is sometimes performed in special cases by organisations with the resources to undertake it, such as the World Bank, but is seldom made public. It is, moreover, not only the avoidance of catastrophes that is relevant, but also the mitigation of lesser but more frequent hardships. For example, a small or medium size reservoir may have no effect in attenuating an extreme flood because the reservoir becomes full to overflowing before the peak intensity of inflow into the reservoir is reached, but the small reservoir may often be capable of attenuating lesser floods nevertheless, which would otherwise have caused costly damage and hardship. This valuable benefit carries with it a well-known managerial and administrative risk that, without proper precautions, downstream inhabitants may be lulled into a false sense of security and become less well prepared and more vulnerable to an extreme event which the reservoir cannot control. Evaluation of reservoirs has to account for the reliability of benefits and the potential risks of losses.

The main alternatives to building more large reservoirs for water supply are groundwater development and/or desalination of salt water, the former being limited and with environmental objections, and the latter being uneconomic except in situations of severe water shortage and entailing acceptable disposal of the salts removed. Power and energy can alternatively be obtained by burning fuel to power turbines or from nuclear power or from renewable sources including solar, wind, wave and tides. There are serious environmental objections to all these alternatives, except perhaps to wave energy, for which, however, no widely feasible design has yet been developed.

Increased agricultural food production, for which the demand for well-regulated water supplies is much greater than for other purposes, can be achieved to some extent by the use and development of higher yielding crop varieties and improved practices, including drainage and the use of fertilisers. The proportion of rain-fed crops can be increased where rainfall permits, but at the expense of utilizing a greater area of land compared with double cropping under irrigation, to give the same yield. Water for irrigation can be obtained directly from rivers or from groundwater but the regulation of supplies to be available when needed throughout the year is sometimes only accomplished by the operation of large reservoirs.

When farmers in developing countries are introduced to irrigation by pilot schemes or staged development, they discover the advantages of regulated water supplies, avoiding crop failures in droughts and allowing increased yields from double and treble cropping. Farmers without the benefit of regulated water supplies for irrigation find themselves left behind and very keen to receive opportunities equal to those enjoyed by the farmers already supplied. With irrigation, necessarily comes improved infrastructure and a general increase in prosperity in the region, along with associated industrial development. Similar considerations apply to the development of improved supplies of power and energy. It is not the case that governments impose unwanted projects with large dams on an objecting population for the benefit of capitalists, but it is the case that governments go to much trouble to raise the finance for large dam projects, in order to satisfy needs and demands and thus to increase their support from the people. The objectives of governments should be thus, but they must ensure that the disbenefits do not outweigh the benefits. In the long run, it is supply and demand that decide what projects are built and what deterrents can be overcome or accepted to allow their building. Experience has shown that it is often very difficult and sometimes impossible for even strong governments to proceed with schemes in the face of determined local opposition. On the other hand, restriction in supplies (and perhaps rationing) implies a discreditable inability or failure to meet demand. The planned hydropower additions over 15 years through to year 2010 in a selection of 14 Asian countries, totalled nearly 135 000 MW (including 16 000 MW in India and 95 000 MW in the People's Republic of China) which amounts to not quite 40% of the total power needed.

It is widely understood that the greatest single cause of human poverty, hunger, deprivation, overcrowding and social unrest is the size of the human population and its massively increasing trend. If attempted restraint on the growth of populations becomes more feasible and successful than it has been so far, the demand for more large reservoirs should become attenuated. It is also the case that reduction of leakage and the more efficient use of water, can postpone or even, in some cases, avoid the need for more large reservoirs. Overall irrigation efficiencies of long standing schemes can commonly be improved from the order of 40% or less to some 60% or more by better management and control — while the water efficiencies of drip irrigation schemes can be of the order of 90%. It is understood that the need for a new large reservoir for water supply in the Thames valley in England was postponed at an early stage for perhaps some five years, by improved detection and reduction of leakage.

Dams and development — the World Commission on Dams

Environmental objections to the construction of new reservoir projects gained exponentially in strength and momentum since the 1950s, causing costly delays and even cancellation in some cases, largely because it became widely realized that projects were being built without sufficient understanding of their long-term effects on the environment and because local inhabitants were unable to avoid losing their homes and livelihoods without acceptable provisions for their future. In the 1970s, Governments and development authorities worldwide increasingly required that the environmental impacts of proposed projects should be assessed and that the assessments should be considered by the authorities and, if necessary, should receive governmental approval before the projects could proceed. As a result of the controversies over the impacts and benefits arising from dam projects, on the initiative of the World Bank and the World Conservation Union, the World Commission on Dams (WCD) was set up to resolve the issues. It was given a 2-year mandate:

(a) to review the development effectiveness of large dams and assess alternatives for water resources and energy development
(b) to develop internationally acceptable criteria, guidelines and standards, where appropriate, for the planning, design, appraisal, construction, operation, monitoring and decommissioning of dams.

This was reported, in *Water Power and Dam Construction* of June 1998, as being welcomed by ICOLD. Twelve Commissioners were selected to represent and assess fairly all the different perspectives in question. Wide-ranging consultation has taken place. A WCD Forum has been established, from 68 institutions in 36 countries, to reflect the diverse range of interests in the dams debate. Substantial funding was provided but effectively only two years were made available for studies and deliberations before producing balanced reports on these vast subjects [6,7]. The reports and recommendations were to be advisory and not legally binding.

WCD began its work in May 1998 and its final report, published by Earthscan Publications Ltd, was launched in London in November 2000 [6,7]. A global and independent review of dams was undertaken, the first component being review of the technical; financial and economic; environmental and social performance of large dams and the performance of their institutional and decision making processes. Eight detailed case studies of large dams were carried out in selected countries together with 17 thematic reviews and a cross check survey of 125 large dams. A total of 947 submissions were received, four regional consultations were held. A second component of the review was an assessment of the alternatives to dams, and a third was an analysis of planning, decision making and compliance issues that underpin the selection, design, construction, operation and decommissioning of dams.

An unprecedented effort was made to avoid possible criticisms that the findings of WCD could be prejudiced or partial. The need for internationally acceptable environmental standards for the development of dams was not and is not in question. It could not be expected that the findings would be acceptable to many protagonists because feelings have run high, beyond the limits of balance and reason, into the realms of emotion. Also, however much care was taken in their preparation, some of the findings will doubtless be subject to criticism and correction (for example, wind power is described as locally and environmentally appropriate and acceptable to the public, despite active environmental objections to such schemes, which include extensive defacement of treasured landscapes). There is clearly the risk that opposite sides in the debate will seize on and quote only the findings and statements of the Commission which suit their own arguments.

The Secretary General of WCD, Achim Steiner, has drawn attention to the importance for developing countries of a few of the key lessons that have been learnt and not ones which developing countries cannot afford. These are: investing in effective options assessment; optimizing the use of existing dams and related water and energy infrastructure; avoiding the potential impoverishment of large numbers of displaced communities and minimizing the potential destruction of environmental assets. Rather, developing countries cannot afford to ignore these lessons.

Those involved in the planning and development of dams are advised to refer to relevant sections of the full report *Dams & Development: A New Framework for Decision Making* [8]. An Overview has been published [9]. There is much detail and innovative thinking but much of the approach is already inherent in good practice and indeed in this Author's and Professor Rao's published works [3–5].

WCD gave the core values, objectives and goals of development which run through the entire report, grouped under five main headings:

- equity
- efficiency
- participatory decision-making
- sustainability
- accountability.

Recommendations for a new policy framework in the report comprise seven strategic priorities and related policy principles, translated into a set of corresponding criteria and guidelines for key decision points in the planning and project cycles, under the following headings.

- Gaining public acceptance
- Comprehensive options assessment
- Addressing existing dams
- Sustaining rivers and livelihoods
- Recognizing entitlements and sharing benefits
- Ensuring compliance
- Sharing rivers for peace, development and security.

A set of 26 guidelines for good practice, occupying 29 pages, is linked to the above seven strategic priorities.

An open and participatory review of all ongoing and planned projects is recommended by:

- using a stakeholder analysis based on recognizing rights and assessing risks, in order to identify a stakeholder forum that is consulted on all issues affecting them
- enabling vulnerable and disadvantaged stakeholder groups to participate in an informed manner
- including a distribution analysis to see who shares the costs and benefits of the project
- developing agreed mitigation and resettlement measures to promote development opportunities and benefit sharing for displaced and adversely affected people
- avoiding, through modified design, any severe and irreversible ecosystem impacts

- providing for an environmental flow requirement, and mitigating or compensating any unavoidable ecosystem impacts
- designing and implementing recourse and compliance mechanisms.

The report includes specific proposals for the following entities to guide them in progressing WCD's proposals:

- national governments and line ministries
- civil society organizations
- the private sector
- bilateral aid agencies and multilateral development banks
- export credit agencies
- inter-governmental organizations
- professional associations
- academic research bodies.

In addition to introductory entries, the report contains 404 pages of text, tables and figures and an early 11-page executive summary. There is a great deal of substance to be digested both by participants and by experience which can be gained in its practice and/or modification. The January 2001 issue of the journal *Water Power & Dam Construction* has published the thoughts on the report of seven participants in the field, including the President of ICOLD, four leading consultants, a major contractor/consultant and James Wolfensohn, President of the World Bank [10,11]. It is generally acknowledged that the report is an excellent work but one which can be criticized as well as praised. Implementation of its recommended procedures is likely to cause delays in development but its recommendations for collaborative processes were already in place before its publication [12]. Its content is intentionally advisory and not mandatory but cannot be ignored by those concerned with the development of dams and reservoirs. It may indeed be made mandatory to some extent by those responsible for promoting developments. It should therefore be perused but not uncritically. The future of dams and development has to be considered, bearing in mind forecasts that by 2025, 1·8 billion people will live in countries or regions with absolute water scarcity. Under these circumstances, more large dams are a necessity.

Environmental impact assessments

The approach to assessing the likely environmental impacts of a project should commence with determining the requirements of the employer in this respect and the regulations and practices effective in the region in which the project is located. Nowadays there are often very detailed regulations published. It is all the more important to confirm to what extent it is acceptable to deal briefly with the subject of impacts which have quite insignificant economic, social or political implications, in order to allow more time and resources to be spent on impacts of considerable importance. When dealing with reservoir projects involving 'large' dams, it is always worthwhile to refer to relevant ICOLD Bulletins, because they cover a wide variety of subjects in some detail, they are prepared by engineers of international standing in their subjects and they carry an encouraging degree of international acceptance, through their approval by representatives of the member countries of ICOLD (which are very numerous and comprehensive). A case in point is ICOLD Bulletin 35, *Dams and the Environment* [13]. This was produced by an ICOLD Committee on Damming and the Environment, appointed in 1972 and replaced by a Committee on the Environment in 1978. It comprises four parts, of which the last is a matrix which is intended to provide a means of listing and evaluating, even if only in qualitative or relative terms, the impact of individual dams and related construction work on specific parts of the environment.

Columns in the matrix deal with the characteristics of actions involved in the possible impacts, with distinctions between the use for which water is destined, the type of action, the zone concerned, physical corrective action and institutional action. The rows deal with the effects on the economic, social, geophysical, hydrological, climatic and biotic environment. Each impact is evaluated under the symbols provided, which introduce the concepts of relative importance, degree of certainty, duration and delayed effects. The completed matrix must always be accompanied by a written commentary, explaining and justifying the user's interpretations.

This Bulletin 35 gives an introduction and instructions for use of the matrix with examples. Under the heading 'General synthesis', it gives descriptions and explanations of the physical, biological and social effects of river development.

Other ICOLD Bulletins [14–17] covering dams and the environment are available:

No. 86, 1992, *Socio-Economic Impacts* [14]
No. 65, 1988, *Case Histories* [15]
No. 50, 1985, *Dams and the Environment* [16]
No. 37, 1981, *Dam Projects and Environmental Success* [17]

It is desirable to carry out a preliminary environmental impact assessment at a very early stage (during conceptual planning or prefeasibility) in order to bring the principal issues into focus and to gain a measure of the studies needed and provisions which may be

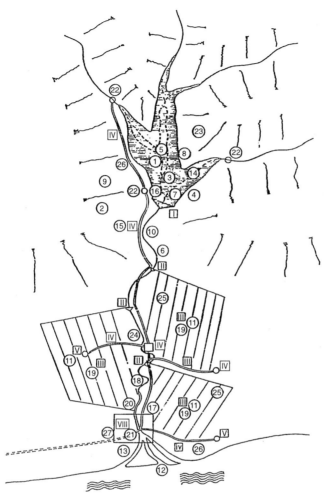

Fig. 13.4. Location of possible effects in relation to siting of a dam

necessary. Bottomley and Clarke proposed a simplified, subjective and qualitative method suitable for this early stage [18]. Figure 13.4 shows a diagram extracted from this paper, illustrating the possible locations of effects in relation to the siting of a dam; a table listing these effects (which are described as 'non-quantifiable') is given in Fig. 13.5. Similar considerations are relevant to the effects of dams for purposes other than irrigation. If the method is followed, effects are listed under the headings: 'Detrimental', 'Beneficial', or 'Detrimental or beneficial according to outlook'. During the later stages of feasibility studies and design, it should be possible to attribute quantitative estimates

232 | Reservoir engineering

	Relative importance scale 0–10	Probability of occurrence (%)	Relative effect scale 0–100
A. Detrimental effects			
1. Human effects of flooding villages	6	100	60
2. Migration of population away from reservoir area	8	20	16
3. Flooding forests and displacing wild life	3	100	30
4. Landslides due to reservoir	2	1	0·2
5. Loss of breeding grounds in lake for migratory fish	3	100	30
6. Reduction of fish breeding grounds in river downstream due to reduced flows	4	50	20
7. Increase of water weeds in lake	3	80	24
8. Unsightly drawdown at lake perimeter	1	100	10
9. Increase in earth tremors	2	5	1
10. Fear of dam failure	8	0·1	0·1
11. Increase of bilharzia	9	10	9
12. Reduced capacity to flush pollutants and salts from estuary	4	100	40
13. Loss of natural fertility due to prevention of flooding and silt deposition	3	100	30
B. Beneficial effects			
14. Development of lake transport	4	50	20
15. Improved transport facilities into area	5	100	50
16. Increase of recreational facilities and amenities	2	50	10
17. Improvement of public services	6	50	54
18. Increased navigation downstream	5	50	25
19. Human effects of settled agriculture and forestry resulting from irrigation	9	100	90
20. Reduction of disease resulting from improved water supply and sanitation	8	80	64
21. Expansion of estuarial port	4	50	20
C. Effects that are detrimental or beneficial according to outlook			
22. Effects of new settlements on displaced persons	9	100	90
23. Change in micro-climate of area	1	50	5
24. Flooding downstream reduced	7	90	63
25. Changes in wildlife, insect population and plant and annual diseases in irrigated areas	7	50	35
26. Introduction of tourism	3	50	15
27. Introduction of new industries	8	70	56
(i) Irrigation dam			
(ii) Diversion dam			
(iii) Irrigated area			
(iv) New or improved roads			
(v) New settlements			
(vi) Market town			
(vii) Port			

Fig. 13.5. The non-quantifiable effects of the irrigation dam (as located in Fig. 13.4)

of lifetime costs or values to all the effects. For this initial qualitative approach, however, they are allocated points for their relative importance, on a scale of 0 to 10. A rather arbitrary probability of occurrence is also given to each, and the product of percentage probability and relative importance marking, divided by 10, gives the position of each on a scale of 0–100. From this it is quite easy for those concerned with promoting the scheme to see which effects are expected to be insignificant and which are likely to deserve detailed consideration.

The fundamental task of an environmental impact assessment is to determine what impacts on the environment will be created by the reservoir project. This is certain to require reconnaissance and survey of the land and its ecology, the natural resources, and the people in the area likely to be affected. The effects of the impacts and ways in which they can be mitigated, or otherwise put to good advantage, are also of fundamental importance for the future of the project, but in some cases their evaluation may be a subject for others to undertake, rather than for those who identified the possible impacts. It is not the intention here to present an exhaustive list of the possible environmental impacts of reservoir projects, which would be endless. Attempts at such lists do exist and need editing for each particular project. Rather, this guide will concentrate on possible impacts of special importance and on their effects, where it is felt that comments ought to be made.

The aspects likely to merit very careful consideration are the following, not necessarily in order of importance; 18 are listed.

- Is the design as good as it should be, resulting in a safe, sound, pleasing project?
- Are sufficient steps to be taken to mitigate adverse effects during construction, such as: disturbance of the countryside, towns and villages by the construction of temporary access roads, by construction traffic, by stockpiling of materials for construction, by construction of temporary camps, workshops and plant yards, by excavation in quarries and borrow pits, by noise, by pollution, by oil spillage, by influx of workers and so forth?
- What measures can and should be taken to preserve sites of special natural beauty, historic or archaeological interest, architectural beauty, etc. from loss due to flooding by the reservoir or due to associated works on infrastructure?
- Will there be losses of natural resources — land, animals, fish, reptiles, vegetation, birds, minerals or oil — that could be mitigated? If so, have such potential losses been fairly valued? (On the basis of supply and demand, a valuation can be made at market

value, if such can be ascertained. Usually there is a market value, and descriptions of features which may be lost as 'irreplaceable' or 'invaluable' have little practicable application.) However, in Chapters 7 and 9 the views of Professor Lafitte were mentioned, sympathetically, that it would not be acceptable to put a monetary value on human life. Nevertheless, this is not the same thing as saying that human life does not have a monetary value or that the value of all lives is the same (regardless of their creative, scientific, commercial, social, political values, etc.).

In some cases, the valuation of losses will entail the valuation of importing replacements for resources lost to the country due to the reservoir, if replacements are needed and do not exist locally. The equation will contain an element needed to value the availability of foreign currency for imports, not always fairly reflected in currency exchange rates. The displacement of wildlife should not necessarily be assumed to be detrimental: animals have shown themselves to be capable of floating and swimming to escape from islands formed as a reservoir fills, and of migrating some distance to establish themselves in a new but compatible environment, if one is to be found. Although this must not be assumed without evidence — and may need help — life forms have an amazing faculty for becoming re-established after they have been displaced. Some ecosystems may be irreversibly lost but new, perhaps incomparable ones may be created.

- Has sufficient been provided for compensation of those who will be displaced by inundation by the reservoir, or displaced by construction of temporary works and infrastructure? Is there sufficient provision for their future, in the form of a living and a lifestyle at least equal to that which they are expected to lose? Persons to be displaced may need new infrastructure and training for a new life. It can easily be argued that there should be a margin in their favour, to compensate them for the disturbance to their lives and for their co-operation in change. However, offering a bonus is a social and political consideration, and not part of the economic cost of the project. Good communications should be established with such people, to explain to them the reasons for the project, what they stand to gain or lose from the project and how their interests are to be protected. Their views must be ascertained and become part of the rationale.
- The dam could cause an impact on migrant species of fish, eels and reptiles by creating an impassible barrier in their path. The existence of species that could suffer from such problems needs to be determined as a result of existing information and/or new surveys and

inquiries. The need for fish passes or lifts to facilitate their passage past the dam, would require study and much has been written on the subject. A possible beneficial impact is natural growth of the fish population in the reservoir, feeding off the flooded land, as well as the possibility of fish being introduced into the reservoir for commercial or sporting purposes.
- Seriously extensive and massive weed growth (notably water hyacinth) has been experienced in new reservoirs, Kariba lake in Zimbabwe being one instance. Usually it has been found that natural processes eventually bring such phenomena under some measure of diminution or control.
- If the lake is to be drawn down to a considerable extent at times, there are likely to be adverse effects on the scenic beauty of the lake and on boat moorings, landings and navigation, round the periphery.
- There can also be increased occurrence of illnesses and diseases round the lake periphery, due to pollution of shallow water by people and due to the creation of breeding grounds for waterborne insects.
- There can be serious effects due to change in the flow regime downstream of the dam, perhaps with diminution in the quantities of water discharged, as a result of diversion of the water by conduit for irrigation or power production, etc. Water levels in such reaches will be lower, perhaps affecting navigation, abstractions and local use of the river water. Reduction in sediment quantity in the water, as a result of sediment deposition taking place in the reservoir, may lead to increased erosion and absence of accretion downstream of the dam. The upstream limits of saline intrusion in an estuary can also be affected, depending on changes in the flow regime in relation to tide levels. In British Columbia, Canada, there have been protests that fish were being killed and spawning affected by untimely releases of water. It has been claimed that water releases from 33 dams do not always coincide with the seasonal needs of the fish and destroy spawn beds. BC Hydro dams have been blamed for hindering the migration of anadromous fish, for creating changes in water temperatures and river channel flow, as well as for increasing mortality of juvenile salmon due to hydroturbines. All such factors need to be accounted for in deciding a minimum daily, seasonal flow regime, which must be maintained by (i.e. 'compensation') water releases from the reservoir.
- Water quality in the reservoir or raised lakes will be affected by lack of oxygenation at depth and by the effects of submergence of vegetable, shrub and tree growth in the area of the reservoir.

There is a usual trend for acidity to increase. These factors need to be studied in relation to the quality of water to be discharged, while the process of discharge itself and turbulent passage of water down the river will tend to restore oxygenation. Water chemistry may also be affected, if filling the reservoir exposes the water to outcrops of soluble rocks, notably limestone, with which it has not previously had contact.

- In the case of creation of a large new water surface, changes in local micro-climate can occur. This is a rather speculative subject, but one requiring study for a sufficiently large project that is subject to local dependence on climate for social or economic reasons.
- The potential for reservoir-induced seismicity (RIE), additional to that which would occur in the absence of the reservoir, is always difficult to assess but can be tackled, based on geology, geomorphology and empirical models. To estimate and quantify the damaging effects of earthquakes, it is necessary to estimate the probability and frequency of intensity of events which could significantly affect the project area, with epicentres possibly up to 125 km distant (for Richter magnitudes 7 or greater, and up to 80 km distant for Richter magnitudes 6 or greater). The potential damage in each case can then be assessed, by relating the potential intensities of events to the local features susceptible to damage. Perhaps the most serious effect of such seismicity can be to trigger slides or slips in slopes in soft ground or rock. Reservoir-induced earthquakes have not been known to exceed a magnitude of about 6·5, which is unlikely to have significant effects at an epicentral distance greater than 100 km. It is desirable to estimate the likely nearest epicentral distance by study of information on faulting in the region and information on past seismic activity. If RIE is thought possible, it is advisable to set up four or five micro-seismic monitoring stations surrounding the proposed site of a major reservoir, as far in advance of reservoir filling as possible. The objective of this would be to determine if filling the reservoir does, in fact, cause an increase in local seismic activity and hence whether the reservoir might or might not be to blame for any earthquake which does occur.
- Instability in the form of slides or slips in soft ground or rock, can also result from rise in pore pressures due to the creation of raised water levels in steep slopes around the new reservoir. Draw down of water level in the reservoir could remove the stabilizing effect of the reservoir water pressure and initiate an instability in the absence of sufficiently rapid drainage of steep slopes.
- If reservoirs are proposed in the vicinity of active volcanoes (the Author has been engaged on one such project in the Philippines

and another in North Sumatra, Indonesia), a major eruption would constitute a major environmental impact on the project. Indeed, the project could be wiped out, as could much of the infrastructure in the region. The loss of the project would probably be a relatively small element in the consequences to the country of such an eruption. If there were a possibility that the creation of the reservoir could lead to flow of water through the rock into the core of the volcano, it might be foreseen that an impact of the reservoir could be the initiation or magnification of an eruption. If such a hazard were to have a significant probability, it would doubtless have to be avoided by changes to the project to eliminate such flow. The monitoring of the activity of active volcanoes, for example by monitoring micro-seismic activity and gravitational variations, is making progress in at least some affected regions. Specialists eminent in the subject of volcanology can be consulted for further guidance.
- Introduction or development of navigation on the lake should be a beneficial impact of the project.
- Improvement of infrastructure in the area should have beneficial impacts.
- Direct benefits from the project in the forms of improved water supplies, improved supplies of power and energy, enhanced agriculture and mitigation of flooding, should all have beneficial direct and indirect impacts locally, as well as providing some increase in permanent employment.

What has been written here on this subject is concluded by emphasizing that care in planning and implementing an environmental impact assessment and in endeavouring to mitigate the disadvantageous impacts and to enhance the beneficial ones, can save untold delays, costs and complications later, and indeed can make all the difference between a successful and appreciated project and one which does not go ahead at all. The Sardar Sarovar (Narmada) water resources project in the States of Madhya Pradesh, Gujarat, Maharashtra and Rajasthan in India is a graphic example of what has been said in this chapter [19,20]. The project was initiated in 1988 but has been long delayed due to environmental, ecological, social and other related issues. Afroz Ahmad [20] has reported the impressive improvements in quality of life that can be achieved as a result of social pressures if the delays and costs arising from their implementation gain the support of government and project promoters and if the extra costs involved do not make the project uneconomic to the extent that it loses the financial support necessary to justify its development (as has so often happened with other projects in the past). This does raise the difficult question of the extent to which different sections of the population are entitled to expect personal improvements in livelihood, the kinds of improvements that can be expected and the sources from which finance for these improvements can be expected, including, perhaps, from the people affected themselves.

The needs for water, advancements in technology and increased understanding by those affected, can be expected to secure the future of reservoirs.

References

[1] Gasser, M. M. and El-Gamal, F. *Aswan High Dam: lessons learned and ongoing research*. International Water Power & Dam Construction, 1994, January, **46**, 35–39.

[2] Cotillon, J. *The High Aswan Dam, A vital achievement fully controlled*. Final review by the Secretary General, ICOLD, Cairo, 1993. November, organised by ENCOLD.

[3] Rao, K. V. and Gosschalk, E. M. *The Financial Case for Hydropower*. The International Journal of Water Power & Dam Construction, 1995, **47**, No. 10, October, 30–34.

[4] Rao, K. V. and Gosschalk, E. M. *The case for impounding reservoirs, an engineer's viewpoint*. International Journal of Hydropower & Dams, 1994, **1**, No. 6, November 1994, 121–125.

[5] Gosschalk, E. M. and Rao, K. V. *Environmental implications — benefits and dis-benefits of new reservoir projects*. British Dam Society, Dams 2000, Bath, June, 199–211. Thomas Telford Ltd, London.

[6] Bridle, R. *World Commission on Dams — The dams debate, London*, Dams & Reservoirs, 1999, **9**, No. 2, July, 22–25.

[7] Binnie, C. *Dams & the environment*. Dams & Reservoirs, 1999, **9**, No. 2, July, 18–21.

[8] Report of the World Commission on Dams. *Dams & Development*. Earthscan Publications Ltd., London, 2000, November, i–xxxviii and 1–404.

[9] Ibid. *An Overview*. Ibid., London, 2000, November, 1–28.

[10] Varma, C. V. J. *Testing the waters*. International Water Power and Dam Construction, 2001, **53**, No. 1, January, 18–19.

[11] Wolfensohn, J. *Talking with the World Bank*, International Water Power and Dam Construction, 2001, **53**, No. 1, January, 21.

[12] LaBolle, L. *Collaborative relicensing: can it work?* Water Power & Dam Construction, 1999, **51**, No. 6, June, 25.

[13] ICOLD (eds). *Dams & the Environment*. Bulletin No. 35, International Commission on Large Dams, Committee on the Environment, 1980, June.

[14] ICOLD (eds). *Socio-economic impacts*. Bulletin No. 86, ibid, 1992.

[15] ICOLD (eds). *Case histories*. Bulletin No. 65, ibid, 1988.

[16] ICOLD (eds). *Dams & the environment*. Bulletin No. 50, ibid, 1985.

[17] ICOLD (eds). *Dam projects & environmental success*. Bulletin No. 37, ibid, 1981.

[18] Bottomley, A. and Clarke, C. L. *Evaluation of the effects of irrigation dams in developing countries*. International Commission on Large Dams, 1973, Congress, Madrid, Spain, Q40, R.39.

[19] Morse, B. and Burger, T. R. *Sardar Sarovar — The Report of the Independent Review*. Resource Futures International (RFI) Inc, 1992, Ottawa.

[20] Afroz Ahmad. *Environmental & social impacts associated with Sardar Sarovar (Narmada) water resources project, India: a success story of conflict resolution and implementing sustainable development.* Narmada Control Authority, Government of India, Ministry of Water Resources, BG-79, Scheme No. 74-C, Vijay Nagar, Indore-452010, Madhya Pradesh, 2000, 1–10.

Further reading
Rao, K. V. *Prospects of developing pumped storage power from disused mines.* Proceedings of Hydropower into the next Century. 1995, International Journal of Hydropower & Dams.

14. Costs and benefits

General

The subject of costs and benefits is fundamental to the development of reservoir projects, not to mention all other kinds of projects, even those for which cost is not an obvious limitation. As already noted, economics, as practised, is not a method of assessing the justification for a project that is either satisfactory or satisfying to engineers (nor to most other people), but no more acceptable alternative has been found. Autocratic rulers and individuals can and do proceed with investments which they justify subjectively and do not need to justify publicly. Economics is not a subject in which engineers are usually adept nor one which they relish, and it is a subject which has become increasingly complex. Nevertheless, it is a subject of which engineers should all try to have some comprehension, because governments, authorities and employers responsible for promoting developments are usually advised by economists and are reliant on financing agencies which invariably employ economists who, in most cases, therefore, have to be satisfied. It is advisable to try to understand the language spoken by economists in order to be able to speak to them on their own terms. Economics is a controversial subject: economists are inclined to present their findings as if they were incontrovertible but, in practice, they are usually put into question by any other economist who is invited to assess them.

The assessment of the value of a project is most convincingly undertaken by assessing the economic returns possible with and without the project — in other words, what is to be gained by constructing the project compared with the returns which would have been obtained by proceeding with the best of other alternatives. Quite complex and demanding software is available for determining (by iterative optimization) the most economic combination of projects in a sequence and the most economic sequence of development of projects for the production of power and energy, for example. One of the best known and widely used by electricity authorities is WASP (Wien Automatic System Planning Package) [1]. This is a program for determining (by least-cost criteria) power-generating system expansion planning. It is a detailed program and, largely due to the extensive data input required and long running time, it is usually run only occasionally when the output is strictly needed. While it includes detailed routines for evaluating hydroelectric power outputs in relation to reservoir operation, there are reservations about its capability for modelling systems with hydroelectric components. However, the variability of hydrological conditions can be represented in the input data, so that fair value can be given to non-firm energy in contributing to reduction in loss of load probability (LOLP). Another program is WIGPLAN but it is thought that this is less suited to representing hydroelectric components realistically. It may be noted that a project which does not appear in the least-cost power-generating system plan within, say, ten years, will have little chance of being allocated finance in a country's development plan. Reference can be made to *Expression Planning for Electrical Generating Systems — a guidebook*, IAEA, Vienna, 1984.

Cost estimates

Before an economic evaluation can commence, realistic and comprehensive estimates of the expected project costs must be prepared. These must be based on the measurement of the estimated quantities of work to be carried out. In the first place, quantities must be calculated from the design drawings and specifications, on the basis of standard methods of measurement published by the Institution of Civil Engineers. It is important to follow the accepted standard methods (modified only where variations are specified for a particular project) because otherwise the unit cost rates applied to the quantities may not cover all the components included (or else the quantities may not be measured to the appropriate payment lines), and hence the interpretation of the rates by the engineer and the contractor may not be consistent. A major example of the need for consistency is whether the measurement or the cost rate (but not both) should include unavoidable overbreak of excavation, over and above the net volume required to fulfil the design and, in either case, how much overbreak is to be measured, or alternatively allowed for in the rates. When preparing documents for tendering, before measurements are taken off the drawings and quantities are calculated, in order to avoid abortive work arising from mistaken assumptions it is advisable that the relevant general conditions of contract, technical specifications, preamble to the Bills of Quantities and variations from standard conditions, should have been at least drafted and compiled. For prefeasibility and feasibility estimates, such conditions will not have been drafted but those assumed in the estimates need to be made clear for future reference and for checking the estimated total cost.

The accuracy expected of cost estimates for hydropower projects should be of the order:

prefeasibility stage:	within ±40 to 50%
feasibility stage:	within ±15 to 25%
pre-tender stage:	±10%
when all tenders received:	±5%.

Clearly, there has to be a progressive process of improving and refining cost estimates as studies and design proceed. In Chapter 3, some guidance was given on the preparation of cost estimates during the prefeasibility stage, where it was stated that the estimates can be prepared by experience and comparison with analogous schemes. It was noted that adjustments should be made to unit rates obtained from other schemes to account for conditions in the other schemes which are different, e.g. for varying freight and transport costs, exchange rate variations in different countries, price escalation if there has been a time lapse between estimates, etc. The importance was noted of ensuring that sufficient allowance in the rates is included for contractors' overheads and profits. A mark-up on the estimated total of direct costs needs to be added for general items, which could include the provision of performance bond, contractors' third party and general insurance, temporary construction camps, workshops, plant yard, storage and materials processing and handling facilities, and access for construction. An allowance must be included for temporary works such as river diversion. An addition needs to be made for unbilled items, that is to say minor items that are too insignificant to bill separately at an early stage. An addition needs to be made for unforeseen contingencies, which should be of the order of the possible positive margin in the accuracy of the estimates at each stage, indicated above. An estimate of the likely cost of escalation (inflation) of the project cost until completion is sometimes required but, more usually, estimates are fixed at the specified date at which they have to be presented. In the case of oil prices, forecasts of escalation may be speculative, but nevertheless important: forecasts of crude oil prices are usually assessed and recommended by the World Bank.

Financing agencies that sponsor foreign consulting or contracting services in a country, by giving free aid or by making a loan on concessionary terms, frequently decline to allow the aid or loan to cover tax on payments from it, if charged by the recipient country, for example on income of the foreign firms or foreign personnel paid from the aid or loan (since this would entail the financing agencies paying more than the direct costs which they intended to meet). In these circumstances, the foreign firms would expect to pay only taxes levied by the governments of their own countries, clearly eliminating double taxation. The point here is to ensure that unit rates and prices do not include for local taxes where they would not be charged on the project — although it is necessary to be sure what is agreed between the donor agency and the local government in this respect.

Despite what has just been said, it is justifiable to include in the cost of fossil fuel a premium, perhaps in the form of intended taxation, to allow for the costs to the country concerned of restraining fossil fuel consumption and restraining the pollution and CO_2 emissions which this consumption entails. If such a premium is not included, comparisons between the net present value of fuel-burning projects and hydropower or other renewable energy projects overvalue the net present value of fuel-burning projects. The premium on fossil fuel has been put at 2p per kW h in the UK. The Royal Commission on Environmental Pollution [2], which reported in October 1994, recommended that pollution should be contained by, *inter alia*, doubling the price of petrol in ten years, which is equivalent to an average rate of escalation of just over 7%, while in some developing countries, it has been estimated that the cost of fossil fuel used in estimating the cost of a project should be increased by some 20% to allow for the environmental on-costs. To finance programmes for control of CO_2, several countries have introduced or are planning to introduce a carbon tax. This, and the costs attributed to CO_2 emissions, have been reviewed by Oud [3]. In the not-far distant past, such factors in favour of hydropower schemes have been disregarded or overlooked.

Unless specifically excluded, an item should be included in the estimates, for estimated interest during construction, that is, interest on the finance which has to be drawn during construction of the project, to make payments to the contractor(s), before financial returns from the project are available for this purpose. Finance for payments to the engineer is usually obtained under separate arrangements, unless the work is to be undertaken under a composite contract for finance, design and build. The engineering costs, including for design, preparation of tender documents and supervision of construction, including expenses, are, however, part of both the economic and financial costs of a project and can be expected to add between 15 and 20% of the total direct costs to the total direct costs, the latter including the allowances for unmeasured items and contingencies.

The mark-ups for some of the additions to be included in cost estimates noted above might be indicated as shown in Table 14.1.

When a price is required for an item, costs quoted for the same item at different dates can be plotted against the dates when they were applicable and the curve best fitting the points plotted can then be extrapolated to the date required. Judgement needs to be used in putting weight on

Costs and benefits | 245

Table 14.1. Mark-up of additions to be included in cost estimates

Addition	Mark-up
Contractors, overheads and profits	25–100%
General items	10% (better estimated individually)
Unbilled items	10–15% on total direct cost
Contingencies	Prefeasibility stage 40–50% feasibility stage 15–25 % pre-tender 10%, tenders received 5%
Engineering	Including design and supervision of construction, 15–20% on gross total estimated cost (the higher limit being applicable to projects outside the UK)
Escalation (inflation)	Can be estimated from UK construction cost indices published by the Institution of Civil Engineers (E. C. Harris/*NCE Index*, January 1989: 100) [4] or from the US Bureau of Reclamation cost indices for water and power projects in 17 Western States, published by *Engineering News Record* quarterly in January, April, July and October [5]. The January 1987 prices are given in Table 6.4 of Gulliver and Arndt's *Hydropower Engineering Handbook* [6] (see reference [13] in Chapter 7) (see Fig. 14.1) and the USBR indices at that date are given in their Table 6.5. (Base date 1977, since when, up to 1987, in 10 years, gross inflation shown has varied from 27% for earth dam structures to 81% for roads, see Fig.14.2.)

Item	Price, USCS units	Price, SI units
Land clearing		
Light	$2000/acre	$810/hectare
Heavy	$5000/acre	$2025/hectare
Earthwork		
Earth excavation	$5·00/yd^3	$6·55/m^3
Rock excavation	$25·00/yd^3	$32·75/m^3
Backfill	$10·00/yd^3	$13·10/m^3
Concrete, moderately reinforced	$375/yd^3	$490/m^3
Penstock steel	$1·50/lb	$3·30/kg
Access road		
Single-lane, paved, new	$145 000/mi	$90 000/km
Two-lane, paved, new	$295 000/mi	$185 000/km
Single-lane, unpaved, new	$90 000/mi	$56 000/km
Existing road, upgrading	$50 000/mi	$31 000/km
Bridge		
Prestressed I-girder type, new	$75/ft^2	$7.00/m^2

Fig. 14.1. Generalized unit prices (January 1987 US$)

Construction item	Cost index
Composite index	1·60
Dams	
Earth	1·41
Structures	1·27
Spillway	1·51
Outlet works	1·61
Concrete	1·60
Steel penstocks	1·65
Canals	1·50
Earthwork	1·49
Structures	1·54
Conduits (tunnels, concrete-lined)	1·70
Power plants, hydroelectric	
Building and equipment	1·66
Structure, reinforced concrete and improvements	1·55
Equipment	1·72
Turbine and generators	1·74
Accessory electrical and miscellaneous equipment	1·63
Pipeline, concrete	1·64
Switchyards	1·60
Transmission lines	
Wood poles, 115 kV	1·47
Steel tower, 230 kV	1·70
Roads	
Secondary	1·81
Bridges, steel	1·64

Base: 1977 = 1·0

Fig. 14.2. *Construction cost indices of the US Bureau of Reclamation, January 1987*

the more recent of the comparable costs, and on those with the greatest similarity in work covered, to the work being priced.

The method of cost estimating using unit rates, just outlined, can be used with increasing refinement and detail for all the stages of prefeasibility and design. At the prefeasibility stage, it can be supplemented and/or checked by estimates based on the cost of comparable major components (or even of complete projects) for which the costs are known in other projects (see *Hydropower Engineering Handbook* and reference [7]). Clearly, such costs have to be adjusted for content, escalation and location, on the lines described for preparing unit rates.

Another approach to cost estimating, largely independent of those so far described, is to estimate the kinds of personnel, skilled and unskilled labour, plant and materials required to construct the project, with the times for which each has to be mobilized. Appropriate hourly or daily unit rates for each kind of resource have to be determined and applied, to obtain an overall estimate of the cost of the project. The method, of course, entails working to a carefully prepared, realistic programme, which makes reasonable allowances for factors such as time lost due to seasonal high river flows and rainfall, maintenance of plant and breakdowns, holidays, etc. It is necessary to be sure that the additional items listed previously are covered.

Principles of project appraisal

Some of the parameters and terms used in project appraisal ought to be explained and understood. Some 15 principal terms are explained in the remainder of this chapter.

Present value

This measure of value of future costs and returns has been referred to in Chapter 3: it is the amount that would need to be invested or would be received at a given date (usually near the date of presentation of the estimates) to yield the actual cost(s) or return(s) when they occur in the future. The investment is assumed to earn compound interest at a particular rate, usually referred to as the discount rate. The net present value is the present value of receipts less the present value of costs.

Opportunity cost

To be acceptable, a project must earn a return equal to or greater than that required by investors on their money. The project must also earn at least as great a return as any alternative project which could take its place. The opportunity cost is the rate of return which could be earned by the alternative project on the investment which it would need, or the rate of return which has to be offered to attract external finance (similar to the test discount rate which is to be described later). In either case, the net present value of the project proposed, if economically viable, needs to be positive when calculated at a discount rate equal to the opportunity cost.

In many cases, hydropower projects have been described as 'not feasible' on the basis of conventional economic criteria, which attribute little value to the present discounted benefits earned after a life of some 20 years, and attribute little present cost to the costs incurred after such a time. (The present value of £1 twenty years hence, when discounted at 10% per annum, for example, is only 15p.) These criteria tend to belittle some of the main attractions of hydropower schemes, namely their long life and low recurrent costs, whereas their high initial capital costs and interest charges on capital seem to show them at a disadvantage compared with power and energy generation schemes using fossil fuel, despite the unending fuel requirements of the latter at escalating cost and their sacrificial use of natural resources. No equally satisfactory alternative has been found to the use of monetary values and economic criteria for evaluating projects. Some, including the present Author, however, consider that more acceptable comparisons can be made if the internal rates of return looked for from hydropower schemes are reduced by a figure (probably of the order of 4%) to offset the effects of inflation, if inflation is not charged on the variable recurrent costs of alternative fuel-burning schemes.

Shadow prices

Shadow prices are calculated in an attempt to adjust the market price to reflect the real cost to the country. If there is an economic advantage to the country in utilizing local materials or labour rather than imported equivalents, the shadow price of the items will be a fraction of the market price, in order to reflect the savings to the government. Similarly, the shadow price could be greater than the market price, if there are costs to the country not included in the market price. The most widely used authority on, and source of shadow prices, is the International Bank for Reconstruction and Development (IBRD), usually known as the World Bank. There seems to be a certain mystique about the calculation of shadow prices.

Internal rate of return

The internal rate of return is the discount rate which would result in a net present value of zero. If present values are calculated using financial (market) prices and values, the parameter is known as the *financial internal rate of return (FIRR)*, while if calculated using economic prices and values, it is known as the *economic internal rate of return (EIRR)*.

Pay back period

The pay back period is the period (usually measured in years) required for the cumulative net present value of the returns on the project, to equal the present value of finance invested in the project to that date.

Sinking fund

The sinking fund is an equal regular annual investment at compound interest, to redeem the loan or investment in the project, or to prepare funds for replacement of major components of the project at the end of their economic lives. This process is sometimes referred to as *amortization*. Care needs to be taken in making the provision because, due to escalation, the cost of the replacement may be different from the original cost.

The period for amortization of heavy civil engineering works including dams, has customarily been taken as 40 to 100 years. At an interest rate of 8%, say, the sinking fund would require only 0·174% p.a. of the sum required, if remaining at a constant price, for redemption after 50 years. In the case of electrical and mechanical plant and equipment, a period of amortization of 25 to 30 years would be customary and, at 8%, the sinking fund would require 1·37% p.a. for redemption after 25 years. If the interest rate were 12%, the figures would become 0·04% for 50 years and 0·75% for 25 years. The sinking fund is nowadays usually presumed to be set aside from revenue. It can be seen that the cost of provision for replacement is not a major factor if the life of the asset is very long and the rate of interest is relatively high.

Operation and maintenance costs (O and M costs)

Operation and maintenance involves the overhead of employing trained staff with the related office and administration costs, as well as the cost of spare parts. Cost estimates are best based on reliable data from local past experience in similar cases. As a rough guide, average annual maintenance charges for heavy civil engineering works have been thought of as about 0·1 to 0·2% of capital cost, while for stationary mechanical plant, the estimate is increased to 2·5–5·0% and for stationary electrical plant to 2·0–4·0%. Down time for repairs and overhaul of E and M equipment needs to be allowed for. In the absence of specific data, it is reasonable to allow, say, an average of about two separate periods of one week each per year and up to two months for an overhaul every four years, or an average of about one month per year down time, allowing for inspections. Running costs should be based on specific related experience. For hydropower projects, they are relatively low but must include the cost of services required and maintaining the availability of operating staff (or a share of remote operation costs for stations which are partly unmanned).

Operating costs should normally be charged against revenue, but whether maintenance costs should be charged against revenue or capital is open to argument. The important consideration is that they should be charged against one or the other.

Losses

Losses in irrigation projects will be dealt with in Chapter 15. The output of a power plant is usually estimated at the outward connection of switch yard to transmission line. Losses in the transmission and distribution system have to be allowed for in estimating the value of energy utilized. It has been noted that these should not be allowed to exceed about 10%.

Assuming that the input to the project is the estimated runoff from rainfall into the reservoir, the gross losses in a hydropower project, starting at the upstream end, commence with open water evaporation from the reservoir surface and leakage losses from the reservoir, both of which should be accounted for in the reservoir operation simulation model. The losses in the power conduit comprise the following.

(a) Losses through the intake screens, largely a function of the square of the velocity through the bars or mesh allowing for partial blockage of the screens by trash.
(b) Losses through the intake, which can be expressed as a function of the velocity head after the contraction of the waterway, the function depending on the streamlining or bell-mouthed design of the intake.
(c) Losses at transitions, which can generally be expressed as a function of the velocity head at the smaller end of the contraction or enlargement. It should be kept in mind that the size and hence the cost of gates and valves can be reduced by transitional reductions in the conduit diameter upstream and transitional reinstatement of diameter downstream. Ideally, the angle of divergence of the perimeter from the parallel, should be around 1 in 10 (too sharp a divergence does not achieve sufficient reduction in losses while too gradual a divergence causes unnecessary costs in construction).
(d) Losses across slots for screens or gates.
(e) Losses through gates or valves. The discharge coefficient of a gate or valve is the discharge through the gate or valve, expressed as a fraction of the free discharge, as it would be if there were no losses due to friction, turbulence or restrictions in cross section.
(f) Losses at bi- or tri-furcations in the conduit, dependent on the diameters, roughness and geometry.
(g) Losses at bends, which can generally be expressed as a function of the ratio of radius of bend to conduit diameter, and a function of the angle of intersection at the bend.
(h) Losses throughout the length of the conduit, which can generally be expressed as a function of velocity head, length, roughness and diameter, or similarly by the Manning equation.
(i) Losses at exit from the turbine runner, which are not all recovered in the draft tube enlargement. It is desirable that the turbine and draft tube should be designed and supplied as a unit and that the manufacturer should guarantee the efficiency of the unit to the draft tube outlet. The dimensions of the draft tube affect its cost and the excavation, so the dimensions must be justified. The velocity at the exit from the draft tube should normally not exceed about 1·8 m/s.

(j) Losses at the outfall of the tail race, which may represent a sudden enlargement. These are usually expressed as a function of the change in velocity head, or of the velocity head in the tail race.

It can be seen that the head losses in the total length of power conduit, having been estimated as indicated above, with the aid of published design charts and data, can be represented approximately, for convenience in optimization of diameters, as a function of the velocity head in the conduit. The Gulliver and Arndt *Hydropower Engineering Handbook* [8] states, as a guideline for preliminary design, that the total head loss from intake trash racks to the turbine inlet should not exceed about 3% for base-load units, or 6% for peak-load units. This seems rather arbitrary and over generalized but does provide a yardstick.

The optimum power conduit design and size for a hydroelectric project, must minimize the sum of construction costs, maintenance costs and the value of power and energy which cannot be produced because of head losses. In Chapter 10, maximum permissible velocities to avoid damage to different types of tunnel linings were indicated; these can be used for preliminary sizing of conduits. A preliminary determination of the diameter of a steel or concrete penstock can be determined within $\pm 20\%$, or possibly even $\pm 10\%$, by reference to published correlations, between the diameter of existing penstocks and functions of the rated capacity of turbine, and of net head on the turbine or (in the case of concrete lined conduits) of the discharge. These have been reported in the *Hydropower Engineering Handbook* [9].

Load factor

This is normally taken as the ratio of average power output to the peak power output required to meet power demand. The period over which the factor is calculated should be stated, e.g. year, month, or day of peak demand.

Capacity factor

This is normally the ratio of average power output to installed power capacity (which may include an element for stand-by and/or an element of surplus capacity).

Firm power output

In the case of hydropower projects, this depends on the water which is available for discharge through the turbines at times of high demand. It has sometimes been taken as the discharge which can be made available (with the benefit of reservoir storage, if available) within the range 90 to 98% of the time, in the year with the driest period on record. It can best be obtained from a system simulation model, with the aid of which the acceptability or otherwise of periods of power shortage to meet demand, can be considered. Mittelstadt [10] explained that the dependable capacity of a hydropower project or projects, operating in predominantly thermal power or mixed systems, can be realistically represented by the *average* hydrologic availability of the hydroplant's capacity *in months of peak demand*, because this produces a fair comparison with the loss of load probability (LOLP) of thermal power. Mittelstadt pointed out that (in the US), coal-fired steam plant typically has a forced outage rate of the order of 15% (and hence an average availability of 85%, but in a worst case scenario, it might have no dependable capacity at all, since it could be completely out of service due to a forced outage). This clearly puts the benefits of hydropower generation in a mixed system in a more favourable light. It should be noted that the average hydrologic availability is not the same as the discharge which will be exceeded 50% of the time (it may be less, depending on the shape of the flow duration curve, as modified by reservoir storage). The benefit in capital cost of firm power output can be measured by the saving which it represents in the cost of having to install other new power generating plant (usually thermal but, in the case of small hydropower, often diesel plant) which would otherwise be required to meet the deficit in peak demand.

Firm energy output

This can be described in the same way as firm power output, but replacing 'power' by 'energy'. It should be noted that a hydropower station which can produce the *power* required to meet loads during two relatively short periods of peak demand each day may not be able to satisfy all the demands for *energy* during a year or even a day, because this ability depends on the total volume of water discharged through the station, rather than on the discharge at the maximum intensity of discharge when operating at full capacity. The point is that base load as well as peak load has to be met, otherwise additional generating capacity is needed.

Test discount rate

The test discount rate in economic terms is usually fixed by the financing agency, based on its lending rates. In financial terms, it may be taken as the financial return that can be made on the next best alternative investments open to the prospective owner and should, ideally, be greater than the cost to the owner of borrowing.

Project appraisal

The appraisal of an important reservoir project required by the financing agency can be expected to comprise the following.

(a) Least-cost analysis in economic terms, i.e. analysis to show that the scheme forms part of the least-cost system development programme. The use of WASP and WIGPLAN software for this purpose has already been referred to, in the case of hydropower and energy generation projects.
(b) Cost–benefit analysis in economic terms.
(c) Cost–benefit analysis in financial terms.

For the project to be part of the least-cost development, the present value (PV) costs of the optimum programme including the project should be lower than the PV costs of the lowest cost alternative programme without it, calculated at the specified test discount rate. If this is not the case, then some alternative programme would be expected to produce the required outputs less expensively and should be considered. The discount rate which equalizes (that is, results in equal values of) the present values of costs of the two alternative programmes is referred to as the *equalizing discount rate (EDR)*. All costs that have already been spent or finally committed (*sunk costs*) should be ignored. Costs that are common to all alternative programmes and have the same time of development, may also be ignored.

Economic benefits may be represented by judging the cost of producing the required outputs of the project against the least-cost alternative programme without the project. Financial benefits may be assessed as the revenue from the project. In both cases, the returns on the project are best calculated from a system simulation model which will tailor the outputs that can be utilized to the demand.

The cost–benefit analyses comprise a stream of costs for the project and, in the case of the economic analysis, a stream of benefits which the project is expected to bring to the country as a whole or, in the case of the financial analysis, a stream of financial benefits which the project is expected to bring to the prospective owner. The benefits in the economic stream are based on an assessment using shadow prices and may include savings in foreign currency requirements, through reducing imports or in meeting demands which otherwise could not be met; the benefits in the financial stream are based on actual or proposed tariffs and charges (which include taxes and perhaps subsidies which do not necessarily represent the true cost; from the financial costs, shadow pricing factors are removed or excluded). For the project to be viable in economic or financial terms, the net present value (NPV) should be positive at the test discount rate or, equivalently, the internal rate of return (IRR) should be greater than the test discount rate.

If a project is found to be economically attractive but financially unattractive, a financial subsidy might be warranted or charges or tariffs might be reduced, so that the project would not be abandoned and the economic benefits would not be lost.

Sensitivity analysis

Before the end of an economic analysis, it is usual to undertake one or more sensitivity analyses in order to test changes in the economic and financial viability of a project, which would result from various conceivable changes in parameters assumed in the basic analysis. These changes in parameters might include such factors as:

- delay to, or in, the programme for implementation
- changes in interest rates or test discount rates
- increase (or decrease) in capital costs
- changes to design (or adoption of alternative designs)
- changes in fuel prices
- changes in crop patterns (for irrigation projects)
- changes in crop prices (for irrigation projects)
- changes in any other assumptions.

All of these would affect either costs or benefits and could therefore influence the viability of the project.

References

[1] *WASP (Wien Automatic System Planning Package)*. Developed by Tennesse Valley Authority and Oakridge National Laboratory for International Atomic Energy Agency, Vienna (IAEA), 1980.
[2] The Royal Commission on Environmental Pollution, 18th Report: transport and the environment, October, HMSO, 1994.
[3] Oud, E. (1993). Global warming: a changing climate for Hydro. Water Power and Dam Construction, 1993, November, Vol. 45, pp. 20–23.
[4] Harris, E.C. NCE tender price index which was published by the Institution of Civil Engineers, NCE Index, Thomas Telford Ltd., London, 12th January 1989.
[5] US Bureau of Reclamation Cost Indices for Water and Power Projects for 17 Western States, Engineering News Record quarterly, January, April, July, October.
[6] Gulliver, J. S. and Arndt, R. E. A. (eds.). Hydraulic Conveyance Design for Hydropower Engineering Handbook. McGraw-Hill, Inc., USA, 1991, 1st edition, chpt. 5 5.35–5.41.
[7] Gordon, J. L. Hydropower cost estimates. Water Power and Dam Construction, 1983, May.
[8] Gulliver, J. S. and Arndt, R. E. A. (eds.). Hydraulic Conveyance Design for Hydropower Engineering Handbook. McGraw-Hill, Inc., USA, 1991, 1st edition, chpt. 7 7.3.
[9] Ibid.
[10] Mittelstadt, R. L. Hydro project dependable capacity. Proceedings of the Waterpower 89 Conference, USA. Niagra Falls, 23–25 August 1989, ASCE, New York.

15. Efficient management for irrigation

Introduction

Irrigation produces demands for the greatest volumes of water needing to be met by regulated supplies from reservoirs. In principle, the overall efficiency of irrigation projects can be expressed in terms of the net water requirement of the crops as a fraction of the water diverted at its source, together with the inflow to the crops from rainfall and any project being rejected as uneconomic. This was the case on a major project for some 100 000 ha in Malaysia, where initially, based on past practice, it was assumed that all the crop water requirements would have to be supplied by releases from a storage reservoir. On this basis the project proved unfeasible, while it proved thoroughly satisfactory when planned on the basis of meeting crop requirements as far as possible from effective rainfall and uncontrolled supplies from rivers and streams, and only from the reservoir when necessary to meet shortages in the uncontrolled supplies. The efficient use of the combination of uncontrolled and regulated supplies of course demanded a high calibre of monitoring of requirements and monitoring uncontrolled supplies in order to minimize releases from the reservoir and hence to minimize the size of reservoir needed. This, in turn, demanded good management.

The build-up of demand for irrigated land is progressing relentlessly. This is inevitable in view of the increasing world demand for basic crops and the diminishing amount of good, unused arable land available. The trend is strengthened by the increasing population which requires or prefers a vegetarian diet. The demand is partly served by the use of higher yielding crop varieties and growing more than one crop each year, both of which increase the need for irrigation. Irrigation is the only answer to the uncertain utilization of land, equipment and labour which occurs when cultivation is restricted to rain-fed crops. The role of irrigation is to improve crop yields, to improve their reliability, to allow the cultivation of crops when and where conditions would otherwise be too dry and, if required, to increase the choice of crops which can be grown. It should be the aim to plan, design and achieve the optimum efficiency of any particular irrigation scheme, by benefit cost considerations of the scheme as a whole. A basic outlook, therefore is for an increasing demand for efficiently operated storage reservoirs to provide the controlled supplies of water increasingly needed for irrigation.

The meaning of efficiency

Efficiency cannot be considered in isolation because it is affected by most aspects of planning, design, operation and management. In particular, *inefficiency* may arise through shortcomings in:

- water distribution to the farms
- water distribution within the farms
- losses during transmission
- losses in the fields
- maintenance of canals and drainage
- lack of incentive to save water
- engineering management
- agricultural management
- local behavioural patterns of agricultural communities
- ability and desire of farmers to adopt efficient irrigation techniques.

The importance attached to economics in irrigated agriculture varies widely, depending on the national and the owners' priorities:

- with subsistence agriculture, the cost may have to be a restraining factor, although the farmers in effect work for themselves
- with commercial estates, cost–benefit optimization assumes high priority
- with projects of pressing political importance, costs may have relatively little significance.

Efficiency in supply of water for irrigation can be measured as (a) the quantity of the predetermined water requirement for irrigation reaching the crops as a percentage of (b) the total amount of water supplied from the source or sources of the water. An alternative measure, the value of the crop per unit of water supplied, has a disadvantage in that it does not provide a standard for differing economic conditions and crop varieties.

(a) The quantity of predetermined irrigation water should include, where necessary, water allocated for:
    (i) consumption by the crops to produce the required economic return

(ii) presaturation of the ground if required (e.g. for paddy)
(iii) regulation of the salt content of the soil
(iv) frost control
(v) crop cooling
(vi) losses due to deep percolation, including any amounts needed to maintain groundwater levels.
(b) The total amount of water supplied should include:
(i) the quantity of predetermined irrigation water required ((a) above)
(ii) water required to make good conveyance losses
(iii) water required to make good losses in distribution
(iv) water gained from a water table
(v) rainfall
(vi) uncontrolled supplies from rivers and streams
(vii) water unnecessarily released or pumped, whether avoidably or unavoidably.

Referring to item (b)(v), there is an inevitable waste of rainfall, depending on the timing of falls, intensity, total quantity of rain, water requirement and speed of response in adjustments to the delivery of water supplies, but it is just as important to minimize the waste of rainfall as it is to minimize the waste of other supplies. Efficiency in utilizing rainfall should therefore also be considered separately, where practicable. Similarly, item (b)(vii), oversupply, should be minimized.

It can be seen that the figures for efficiency defined here are indicators of waste, as measured by the difference between denominator and numerator (i.e. (b) − (a)) that is, water lost in conveyance and distribution — the losses which are chiefly within the province of engineering and management to control. They can be minimized as far as it is economic to do so, but the minimum must vary from scheme to scheme, depending on the route from the source of water to the fields and on local factors, namely soil and climate.

Drainage or water seeping from canals or river beds, if subsequently utilized, reduces the quantities of water that must otherwise be provided, and so results in a corresponding improved utilization of water resources as a whole and so, in improved efficiency.

Irrigation efficiency on the lines adopted and described by the Author has the merit, not common to other definitions of efficiency that have been used (which, for example, exclude from predetermined irrigation water all but water consumed by the crops), that it is a measure of the efficiency of engineering and management in achieving what is planned and therefore provides a general basis for comparison of schemes. It must not be overlooked, however, that it is a measure of the designer's success in minimizing losses only when comparing different schemes for utilizing a particular source of water to irrigate particular crops in a particular location, and it does not in itself say anything definitive about economy in cost. Water losses can be minimized by spending more, e.g. on lining canals, but it is clearly necessary to work out an optimum economic balance which is not always in the arrangement that gives the greatest water efficiency. Published figures for the efficiency of water supply in irrigation schemes are often liable to be highly misleading, because of inaccuracies in measurement and uncertainty as to precisely what gains and losses have been taken into account in arriving at the figures. Estimates of efficiency, however, when used intelligently, can serve a useful purpose in comparing different schemes and in making a realistic evaluation of unknown losses on one scheme, by comparison with the known losses of another.

Planning for efficiency and control

There are so many variables that affect the planning of irrigation projects that each project has to be tailored to suit its own particular assets and needs. A conceptual general scheme for the planning and development of irrigation projects is shown in Fig. 15.1, which might be treated as a 'check list'. The best policy is to proceed in progressive, well defined but flexible stages, each of which provides its own opportunities, through the media, of reports and discussions with the authorities and people concerned, for decision making and revision of methods or objectives. In this way abortive work is minimized.

In principle, planning and design should start with the identification of suitable land and a desirable objective, such as the growing of certain viable crops which would benefit the economy and/or serve a known market. The design planner should then proceed in an 'upstream direction' to evaluate the best means of providing irrigation facilities for these crops. Although obviously there are exceptions to this rule, the alternative, of starting with sources of water and 'proceeding downstream' to design developments in order to utilize available water supplies, may fail to disclose alternative and perhaps better means of satisfying the most important objectives.

Once the broad outlines of a scheme are conceived, coherent components can be conveniently separated for investigation, initial outline design and approximate evaluations. The work should proceed by progressive refinement and integration of the parts, until the parts can be knitted together, analysed, refined and evaluated as a whole. A possible simplified sequence of study (flow diagram) is shown in Fig. 15.2. The study may be arranged so that escape from the closed

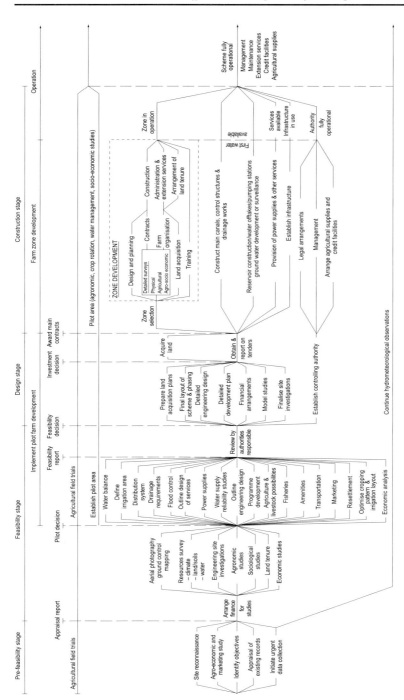

Fig. 15.1. Conceptual analysis for development of irrigation and drainage projects

260 | *Reservoir engineering*

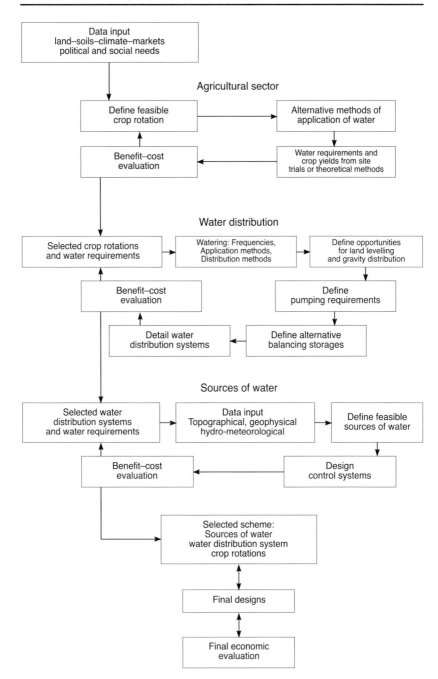

Fig. 15.2. Possible study sequence

loops takes place when successive refinements yield only insignificant improvements.

Ideally, to produce an efficient project, the conception of a scheme to control water allocation and distribution must develop integrally with the project planning and design. Indeed, a reliable estimate of irrigation efficiency presupposes a certain degree of control, and without an appreciation of this, water requirements cannot be calculated within reasonable bounds of accuracy. Nevertheless, control is also the basis on which existing inefficient systems can sometimes be renovated and upgraded within their physical limitations.

Basic criteria

Where an irrigation scheme is deemed possible and needed, first of all it is necessary to confirm suitable crops, depending primarily on:

- soil characteristics
- climate
- potential markets
- costs and benefits.

The investigation of these factors leads to the definition of agronomically feasible crop rotations, each, for the purpose of preliminary planning, having a discrete consumptive use of water. At this stage of conception, it is desirable to make a coarse preliminary cost–benefit evaluation, in order to narrow the options. This involves estimates of water requirements and of the unit cost of water: these can be arrived at with sufficient accuracy for a first approximation, by analogy with other comparable schemes. The outcome is the identification of likely crop rotations, both by extent and season, which allows the designer to proceed to the next step.

Water requirements

It is possible at this point to make a closer estimate of the likely farm water requirements, as a basis for the design of their allocation and the distribution system. The following are the principal factors affecting water requirements:

(a) soil moisture retention
(b) mode of application of water to the land
   (i) by flooding
   (ii) by furrow
   (iii) by sub-surface distribution
   (iv) by sprinkler
   (v) by trickle irrigation
   (vi) by groundwater recharge

(c) climate
(d) type of crops and nature of end product to be utilized
(e) the optimum economic end product (which may not necessarily be the maximum weight or maturity of crops).

The most satisfactory way of establishing water requirements is by local experience — if applicable and reliable — or by prolonged site trials, which are expensive and cannot give early results. One of these two sources of data is desirable, because of inevitable uncertainty about the longer term behaviour of the crops, variability of the soil, of pests and diseases, changes in variety and cultivation methods. In addition, the existence or creation of undergroundwater supplies within reach of deeply penetrating, water-questing roots, would substantially reduce the quantities of water necessary to be applied at the surface. If it is established that crops can in fact draw on water from an existing aquifer, it is necessary to investigate whether this supply could be withdrawn, by natural fluctuations in the water table or by future pumping. In the absence of trials or quantitative experience at the site, preliminary estimates of the water requirements of different crops can be obtained by applying empirical factors to evapotranspiration measured locally (the factors being obtained for different crops from published data) or calculated from records of evaporation or from empirical formulae such as those of Blaney and Criddle [1,2,3], Thornthwaite [4], Hargreaves [5] and others, relating evapotranspiration to humidity, mean monthly temperature, wind speed, pan evaporation and hours of sunshine per day.

Since the application of an increasing amount of water beyond a certain optimum usually yields diminishing returns in the form of increased crop yields, whenever data permit, the optimum net application should be estimated from yield/deficit relationships. When evapotranspiration is less than that required for maximum yield, the range in yield can vary widely for the same total evapotranspiration, depending largely on the stage of growth during which 'water stress' occurs. Timing is therefore important. The range of variation is less as total evapotranspiration approaches that necessary for maximum yield, so there is less risk of incurring drastically reduced yields if a figure approaching the requirement for maximum yield is adopted for design. If sufficient data are available, water requirements may be stated as a function of the state of maturity of the crop and of climatic conditions, for the soils of a particular location. Otherwise approximate average requirements may have to suffice, to be improved in practice.

Potential evapotranspiration is defined as the amount of water used in unit time, by a green actively growing crop which completely covers the soil and which is grown under conditions of non-limiting water supply. Published figures for evapotranspiration should be treated with caution, since they are sometimes seriously in error or misleading, due, for example, to ignoring deep drainage and to inaccuracies in measurement. The more uncertainties there are in estimating water requirements, the greater the water supply which must be made available to guard against underestimates. This of course usually leads to over-design and resulting waste.

Water distribution systems

Some radical design decisions for any irrigation scheme relate to the water distribution system. Key decisions are:

- the frequency of irrigation
- the method of application of water to the land
- the method of delivering water to the farm.

The frequency of irrigation

This may be governed largely by soil moisture retention and infiltration rates. The objective of minimizing the capacity of field off-takes and channels and of minimizing losses by deep percolation, leads to frequent small applications of water. However, the minimum amount of water to be applied in one application must be sufficient to cover the required area rapidly and uniformly and to supply the crops until the next application. The amount is thus a function of plot size, topography and soil conditions, as well as of the type of crop and its root requirements, and the amount is most reliably determined and refined in the course of agricultural trials. If not carried out previously, every effort should be made to set up and carry out trials as part of the first stage of a new irrigation project when it has been completed and commissioned. Because of the variables to be tested and because of climatic changes, for reliable results, trials must continue for at least four years, preferably longer — indeed as long as is feasible.

When sprinkler equipment is to be used, losses in surface travel across the fields are minimized, uniform application is fully controllable, and uniform application is more easily achieved. In the case of portable sprinkler equipment, frequency of application is a function of the convenient timing of moves. In order to minimize equipment costs, water is applied at the highest acceptable rate consistent with water economy. The equipment is then moved to cover as many other areas as possible, before having to return to re-water the same area.

The method of application of water

The method of application of water has been referred to above as a matter affecting water requirements. An auxiliary cost–benefit evaluation is needed to establish, at least as a first shot, whether possible savings in reducing water losses and land unproductively occupied by channels, warrant the adoption of more expensive methods of applying water to crops. At this stage, the question to be resolved is whether the installation of sprinklers is likely to be justified and, if so, how likely. In cases where the existing ground is too irregular for gravity application, the possibility of land levelling has to be assessed.

Sprinklers are an efficient means of reducing the overall water requirement, by directing water accurately where it is needed. It should not be overlooked that, with leafy crops and arid or semi-arid conditions, the savings may be offset by losses due to extensive moist leaf surfaces exposed to evapotranspiration.

The method of delivering water to the farm

Evaporation and seepage losses are minimized by the use of closed conduits, whereas recurrent costs are minimized by reducing pumping and making the greatest practicable use of gravity feed. Whether covered or not, the minimization of seepage losses involves the lining of unlined canals passing through permeable ground, and the effective sealing of joints in the lining. In any case, the properly integrated provision of balancing storages provides the gearing for efficient operation.

Balancing storages

Balancing storages may take several forms:

- surface reservoirs or ponds
- spare capacity in channels
- underground aquifers and porous strata
- storage on the fields in cases where flooding is acceptable.

The provision of storage is not necessarily expensive, and in some cases may incur no extra cost at all. It may serve several valuable functions:

(a) to reduce the canal capacities required upstream and their cost, by allowing water to be delivered to the balancing storage at a low rate over continuous periods, although only acceptable at the fields during limited working hours
(b) to reduce the capacity and cost of pumping plant to be installed at the source of water, by enabling it to deliver water continuously instead of only during farming shifts

(c) to store, until needed, local uncontrolled stream supplies and rainfall which would otherwise be wasted
(d) to provide flexibility in meeting demands which may have changed, due to unexpected rainfall, for example
(e) to provide a reserve in case of prolonged interruption of the main supplies due to breakdowns, maintenance, holidays, etc.

The amount of storage provided is inevitably often assessed arbitrarily, for example to cover 36 hours supply downstream, or to allow the canals upstream to run continuously at an average rate, despite intermittent variable demand downstream. Preliminary assessments can, however, be refined using mathematical methods, which I will return to, thus allowing the capacity of balancing storage to be optimized with respect to the savings the storage produces.

Sources of water supply

The endowments of a particular area for irrigation are its climate, its soils and its water resources. Possible sources of water are the following:

- Rainfall and then, in so far as their quality is suitable:
  - Direct supply from rivers or lakes
  - Impounding reservoirs
  - Underground aquifers
  - Drainage water

The contribution which each source can make depends on the features of the particular area considered. The use which can be made of rainfall can be assessed using a mathematical model, providing that the amounts of rainfall which can be accepted can be defined in relation to the calendar and in relation to the preceding conditions of weather and water supply or in relation to current groundwater levels. Alternatively, average monthly effective rainfall (the proportion of total rainfall which it will be practicable to use effectively) can be estimated from monthly rainfall, monthly consumptive use by crops and depth of application, according to the method developed by the United States Department of Agriculture, Soil Conservation Service.

It should go without saying that the maximum practicable use should be made of rainfall, since this costs nothing to obtain. A similar consideration applies to capillary soil water and groundwater, where this can be reached by the plant roots without pumping. The choice of the more expensive sources of water available to provide the balance of water required is a matter for economic study.

Analysis

The stage has now been reached where the project design can be brought quite sharply into focus. Having decided on promising crop rotations, the method of application and delivery of the water, and on the sources of water supply, the crop rotation and distribution systems can be optimized. This is a suitable problem for linear programming.

Control

The object of control is to enable the system to fulfil its purpose by the efficient and economical use of water. The hydraulic control structures should be designed for flexible control and simple, sufficiently sensitive, measurement of flow, compatible with limited fluctuations in canal water levels. In this respect, overflow gates are superior to undershot types, although sluices for low-level sediment removal may be required as well.

The amount of money worth expending on a control scheme can be judged by valuation of the wastage that would occur without it. The greater the cost of water, inevitably the more it is worth spending on control. At one extreme, there are irrigation schemes where water is distributed in fixed doses according to a basic rota, while at the other, electronic controls are provided to calculate the most economical use of reservoirs, use of water stored on the fields (e.g. during supersaturation of paddy cultivation), use of uncontrolled water supplies and rainfall. A completely automatic system of control of water distribution can be achieved when warranted, possibly involving automatic measurement and transmission of reservoir levels, canal levels, stream off-take levels, rainfall and soil moisture retention, these being input into a numerical hydraulic model on computer. Altogether 33 Reports of the Eighth Congress on Irrigation and Drainage (ICID), Varna, 1972 [6], were devoted to the application of automatic facilities in operation and maintenance of irrigation and drainage systems.

Control is the answer to the problem of delivering water where it is needed and when it is needed. The provision of a forecasting service is a desirable element. The time taken for water to find its way down natural river courses or irrigation channels, and the flexibility provided by balancing storages, are fundamental factors. It must be appreciated that a block of water released from a reservoir at a uniform rate, over, say, 24 hours, will be delivered at a later time, over a longer period, at varying rates attenuated by channel storage. The attenuation might be determined experimentally (empirically), or by a simple graphical method, or by incremental calculation, the latter being convenient for application by computer.

Despite the use of a forecasting service, if water delivery is unnecessarily slow, there may be considerable waste, if unexpected rainfall or uncontrolled supplies occur, while the ordered consignment of water is in transit. Thus liberal provision should be made, where appropriate, for field outlets and field channels of sufficient capacity to speed the distribution of water.

There may also be a delay in ascertaining crop water requirements in a particular area. Where labour is plentiful, the requirements can be observed and reported in writing by inspectors or by the farmers themselves. Where refined equipment is viable, however, the requirements can be conveyed to a control centre by transmitting gauges, registering one or more parameters such as soil tension, evaporation, water level and rainfall. It is, however, necessary to keep the number of gauges to a reasonable minimum, in order to keep the cost within reasonable bounds and to make maintenance manageable. Using the time lags inherent in the system, it is possible to produce a mathematical model to simulate project operation, by which the efficiency and viability of operation can be assessed.

Forecasting

Some degree of forecasting river flows, rainfall and plant requirements is fundamental to irrigation developments and has been employed throughout history, often, perhaps, unwittingly. Such forecasts are implicit, when, for instance, river flow has a certain value and it is judged that it is not likely to change by more than a certain amount in the next 24 hours. Or when it is judged that rain can be expected in the afternoon because there are black clouds about, and it nearly always rains in the afternoon at that time of year. Or in deciding that a certain crop usually takes around so many weeks to mature, and therefore will require water only between certain dates. Such forecasts still have a useful part to play.

With the aid of transmitting gauges and computers, short-term forecasting of river flows has, however, at last become a more precise and practical tool with which to reduce water wastage and so improve project efficiency. The observation of rainfall intensities and extent by radar, if available, can provide very early notice of actual conditions over the catchment and irrigated areas. Forecasting of rainfall more than a day or two in advance is still relatively unreliable. Forecasting likely reservoir inflows can allow better use to be made of reservoir storage capacity (e.g. if a dry period is expected, reservoir levels might not need to be held down to accommodate floods). Forecasting of stream flows likely to come in below a reservoir can allow releases to be made from the reservoir to provide for expected deficiencies downstream, or allow releases to be reduced, when auxiliary supplies are anticipated. The design and provision of an effective forecasting service allows more logical planning and use of balancing storage capacity.

Forecasting of river flows can be carried out by methods based on physical relationships, e.g. using information from transmitting river gauges, sited on tributaries, or by numerical methods, or by a combination of both. In any case it is desirable to assess confidence limits for the forecasts, and to provide a continual and progressive correction of errors. Advantage can be taken of the ability to forecast rainfall, wherever spatial intensities are sufficiently uniform and suitable meteorological services can be made available. With improving meteorological facilities and technology, rainfall forecasting has a most valuable part to play in improving irrigation efficiencies. The question of forecasting crop demands has already been discussed in this chapter. Although, as described above, forecasting is a very valuable tool, the use of automatic controls can modify and reduce the dependence on forecasting by accelerating responses to variations of supply and demand.

Planned water shortages

While the distribution system must be designed to meet the intended demand, it is accepted that the designed optimum supply is neither necessary nor economic at times of severe water shortages. Most crops can tolerate a substantial shortage for a few days at long intervals without serious reduction in yields. Unfortunately, quantitative data on which to make reliable economic judgements usually seem to be lacking. Crop value/water deficit relationships should be used whenever available; in their absence, the use of Beard's Shortage Index [7] appeals to common sense. It is defined as the expected sum of the squares of all annual shortages during a 100-year period, if each shortage is expressed as a ratio of firm yield. This implies that shortages of 40% are four times as serious as shortages of 20% and that a shortage of 60% would be nine times as costly as a shortage of 20%. It is considered to be a practical index for ordinary planning purposes. Values of Beard's Index, of the order of 0·25 (corresponding to a single shortage of 50% in 100 years, or 25 shortages of 10% each in 100 years) appear to be often acceptable. Where possible, to sustain confidence, as well as to minimize the effect on crops, unavoidable shortages should take the form of a reasonable reduction of supplies over a longer period, rather than complete withdrawal of supplies over a shorter period.

Final design and economic analysis

Having arrived at an appropriate efficiency of water supply for the scheme, the stage is set to complete the design, and to carry out a full economic analysis, which could still lead to final improvements. However, the last words here on this subject are given to caution: temptation to continue revisions to design at a late stage, before as well as after contracts have been let, should nearly always be resisted. The superficial advantages of late design changes frequently turn sour in the face of the abortive work and delays (and increases in cost) that are created by late design changes. A good designer, like a good traveller, is one who can find his way without continually changing direction.

References

[1] Blaney, H. F. Climates as an index of irrigation needs. *Yearbook of agriculture 1955*, US Department of Agriculture.
[2] Criddle, W. D. Consumptive use of water. A symposium. Transaction of American Society of Civil Engineers. Vol. 117, 1952.
[3] Blaney, H. F. and Criddle W. D. Determining consumption use for water developments, irrigation, and drainage speciality conference. American Society of Civil Engineers. Las Vegas, Nevada, 1966, November.
[4] Thornthwaite, C. W. An approach towards a rational classification of climate. *Geological Review*. Vol. 38, 1948. 55–94.
[5] Hargreaves, G. H., *Consumptive use derived from evaporation path data*. Proc. American Society of Civil Engineers, Irrigation and Drainage Division. Vol. 94, IR 1, 1968, March.
[6] International Commission for Irrigation and Drainage (ICID) (eds.). Eighth Congress. Q 28-2, 33 Reports on application of atomic facilities in operation and maintenance of irrigation and drainage systems.
[7] Beard, L. R. *Estimating long-term storage requirements and firm yields of rivers*. General Assembly at Berkley of I.U.G.G. Symposium, Surface Waters. IAHS Publication No. 63, Gentbrugge, Belgium, 1966.

16. Small hydropower

Introduction

There has been, and still is, some confusion over the classification of 'small' hydroelectric schemes. Definitions reported by United Nations International Development Organisation (UNIDO) were:

micro: less than 100 kW
mini: 101 kW to 1 MW
small: over 1 MW to 10 MW

However, 'small' has been classed in the USA by some as being less than 15 MW, *or* as having dam heights of less than 20 m. Elsewhere it has been classed as being less than 5 MW with runner diameters less than 2 m. In the Philippines, schemes of total installed capacity of up to 50 MW were described as 'small' in the early 1990s. In principle, these guidelines consider 'small' as over 1 MW to 10 MW.

International interest and growth in the development of small-scale hydroelectric schemes has continued to grow substantially since the 1970s, considerably stimulated by a series of international conferences sponsored by the magazine *Water Power and Dam Construction*. However, by 1984, China already had over 9000 MW capacity installed in small schemes, of which 95% were less than 500 kW. The interest has centred not only in developing countries but also in the developed world (e.g. in the USA, Norway, France, Italy, Greece and in the Czech Republic and Slovakia). Little encouragement has been given to the development of small schemes in the UK, in fact rather the reverse. The role of small hydropower is different things to different people and to different countries. To one electricity authority it may be an unwanted nuisance, to another, an irreplaceable means of meeting unfulfilled demand, especially in remote areas. To one country it may be a saver of imported fuel, to another an expensive user of scarce capital. The advantages seen in the implementation of small hydropower can be summarized as follows.

- As for medium and large hydropower, once built, energy generation is almost immune from inflation (and pollution).

- If well planned and engineered, the capital cost per kilowatt installed and per kilowatt hour generated *can* be as low as or lower than that of larger schemes.
- Small schemes take far less time than large ones to plan, finance and build and hence the investment can far more easily be tailored to energy demand, and to changing forecasts of energy demand.
- The siting of schemes is relatively flexible, and there are more opportunities to site them strategically close to points of demand.

Having said that, it needs to be said that both the technical and economic problems to be overcome in implementing small schemes are similar in nature to those of larger schemes. While problems can be equally difficult, there are far smaller financial and time resources available to deal with them. A major conclusion is that, preferably, small schemes should not be developed in isolation but in a balanced sequence, allowing the economics of sharing all the following between more than one scheme: access, transmission, construction management, mobilization, demobilization, operation, maintenance and engineering and also allowing repetition (as far as reasonable and sound) of design and manufacture.

Some criticisms of small hydropower can be summarized as follows.

- Schemes have been developed where the cost has proved excessive or which have failed due to poor engineering, e.g. due to unsuitable choice of plant, or intakes destroyed by floods or filled with sediment, or pipelines and roads affected by landslides, etc.
- In a large electricity authority, small schemes may not receive the necessary managerial attention because of the trouble of integrating them into the system and the problems of operation, maintenance and repair, with only a very small contribution to the system's generating capacity in return.
- Small schemes may become superseded when larger schemes are built.
- Small hydropower is not an economic means of providing power and energy to meet major concentrated demands, largely because of the economies of large scale and the high cost of transmission from scattered and diverse sources.

Criteria for successful development

A scheme cannot be described as successful if it is not competitive when compared to alternatives. To determine this, an assessment must be made involving comparison of the cost of the small hydro project or projects in question with:

- provision of power and energy from other sources and
- use of alternative types of energy at the demand centre and

- use of the capital required for some other purpose altogether.

The assessments can be made by the conventional methods used for larger schemes. The World Bank has issued guidelines for such assessments [1,2].

Means of siting and planning hydropower schemes in a river basin were outlined in these guidelines in describing prefeasibility studies in Chapter 3.

Sound planning and design

To avoid problems that have prejudiced the success of past schemes, careful attention needs to be paid to the following.

(a) A correct assessment of water availability must be made. There is liable to be great variability in rainfall, while runoff on small catchments is particularly vulnerable to the effects of change in land use. As an example of the worst sort of planning, streams which were assumed to be perennial have proved to be only seasonal.

(b) A correct choice must be made of plant capacity and type. The rules are quite well known but not always followed. Important aspects are: making an economic choice to meet peak load requirements, taking account of water availability; suiting plant characteristics to the probability of variations in head and in discharge. The choice between synchronous or induction generators is important, with the cost advantage going to the latter if feeding into a sufficiently large system to provide frequency control. The provision, where economic and necessary, of speed increasers between turbine and generator, as well as governing and load control are important considerations. Judgements on the most suitable number of turbine-generators and on the necessity for stand-by plant, must be made. Electro-mechanical protection against under- and over-load and faults must be incorporated.

Machine design should be robust, but not over-simplified (which could lead to unreliable operation). For example, minor savings on protection equipment and on small but critical details, are certain to be false economies. The design must pay full regard to ease of maintenance, sometimes carried out by semi-skilled staff. The correct choice of plant must extend to the correct choice of facilities throughout the project.

(c) Adequate assessments must be made of ground conditions in order to avoid disruption of access roads, conduits and other facilities by landslides or foundation failures.

(d) Adequate estimates must be made of the rate of sedimentation at intakes, and of the floods which need to be passed. The works must be designed to withstand damage from debris brought down during floods. Adequate means of dealing with sediment, including boulders, and for passing floods safely, must be provided. It may be cost-effective at the design stage to provide bottom outlets, perhaps with flexible pipe intakes for the discharge of silt from the reservoir bed, rather than wait for siltation problems to develop, with no economic solution available, bringing the probability of having to cease operation. The concept of flexible pipes connected to bottom outlets through the dam involves the pipes being supported by pontoons on the reservoir surface and used for a form of suction dredging without pumping, using the head in the reservoir to discharge, below the dam, reservoir water laden with sediment. It is necessary to determine how the sediment discharged can be accommodated acceptably, wherever it comes to rest, downsteam of the dam.

(e) Adequate provisions must be made for manageable and satisfactory operation, maintenance and repair. If possible, a central organisation to perform at least some of these functions for a group of small hydro schemes has obvious advantages. Robust and simple equipment not prone to faults is called for. Sensible consideration must be given as to whether the extra cost of automatic or remote operation is justified. As a rough guideline, remote control equipment for turbine-generators might be expected to add, say, £17 000 to the capital cost per unit.

Multiple development

The development of a significant number of small hydropower schemes in series or in parallel with the economies of shared facilities, manufacturing resources and engineering, can have a decisive advantage over the development of schemes in isolation. The development of a complex of schemes would preferably take place in the same or adjoining river valley(s), in order to gain the advantages of sharing the cost of:

- access roads
- transmission lines
- mobilization and demobilization of work force and construction plant
- operation and maintenance
- engineering.

The possibility of economy through repetition of design and manufacture may be realized in this way, together with economy arising from continuity of work by manufacturers and contractors. The multiple development of several schemes together increases the financial commitment involved, and this has one major advantage — that it justifies the sound planning, engineering and supervision and construction management that are so necessary to ensure a satisfactory outcome and to protect the investment.

Operation at optimum load factor

At first sight, a scheme can be economically justified most easily if it is to operate at the highest possible load factor, that is to say, as continuously as possible, at nearly average output: the number of units to be installed and the cost per unit is minimized, for example, in the case of an induction generator tapped into the grid, i.e. a so-called 'grid tie' scheme, where all the energy that can be generated can be utilized at high load factor. Even when feeding a grid, however, particularly a grid which has a thermal or nuclear input to meet base load, there can be overall benefits to the system by installing a larger capacity of small hydro to contribute to peak requirements. For instance, in the UK it is now considered that it would have been advantageous if hydropower schemes could have been built to work at less than half

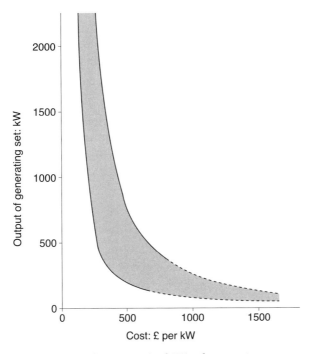

Fig. 16.1. Cost envelopes for output (in kW) of generating set

the load factors adopted some 35 to 45 years ago. The extra value of the power at peak loads more than compensates for the increased size and cost of the plant.

To quote a World Bank publication [3] in 1980:

> ... their relatively high investment costs (illustrated in Fig. 16.1) may make mini-hydro projects (under 1 MW) uneconomical for village systems with low load factors. If they can be connected to a central grid, mini-hydro plants can be more effectively used to replace energy supplied by fuel, as well as to provide additional capacity. In isolated areas, for village electrification schemes and small industry, they can also supplement or replace high speed diesels. However, development of this source is likely to be held back by the scarcity of managerial and engineering talent.

The International Bank for Reconstruction and Development — World Bank (IBRD) [4] estimated that mini-hydro projects of under 1 MW capacity might comprise 5–10% of the world's total hydro resources.

Conjunctive operation

Conjunctive operation with other small hydro or other energy sources may be the key to economic fulfilment of demand. Similarly, the combination of connecting other sources and the staggering of loads, if possible, may help to produce a higher load factor and a satisfactory demand curve, which is compatible with potential supply.

The needs for small hydropower

The needs for small hydropower usually fall into the following categories:

(a) utilization of a renewable natural resource
(b) providing flexibility in meeting demand
(c) where no competitive alternative is available
(d) the provision of relatively small amounts of power and energy at modest capital cost.

Utilization of a renewable natural resource

Since river flow is a natural asset, there is intrinsic merit in utilizing such an asset *per se* for the greatest advantage to the nation to which it belongs. This is really a matter of good housekeeping, and should not be unduly influenced by transient economics, providing that finance can be found and that the energy generated can foreseeably be utilized. If finance is severely limited, it must of course be asked whether the country has more urgent basic projects that must be given priority. While changing economic factors sometimes seem to favour hydropower and sometimes do not, it has been demonstrated over the last century that hydropower, which is relatively inflation-proof, is a continuingly desired facility that benefits users over the years. One could liken it to building one's own house compared with the alternative of renting accommodation; in the UK, the family's own house is often its most important asset.

Whether a small or large hydropower project should be built is a matter of what can be afforded and what is needed. Properly designed and implemented, the capital cost of small schemes per kilowatt installed can compete favourably with that of large schemes. The recurrent annual cost per kW h generated is a better measure of the financial value of a hydropower scheme, but neither of these parameters tells one anything about the peak demand for which the station was designed or the load factor at which the energy is to be generated. They are not very helpful in comparing the merits of different schemes, unless the load factors and peak demands for the schemes compared are known and taken into account. It will be shown presently that there is often a case for building both large and small schemes in series, to fit projected demand.

The utilization of a renewable natural resource has the automatic indirect benefit of reducing dependence on recurrent imports of the fuel that might otherwise be necessary, and/or of reducing consumption of a valuable resource such as wood fuel, oil, gas, coal, etc.

Providing flexibility in matching capacity to demand

Matching power and energy generation to demand is not only a matter of timing but of location. The siting of small schemes is more flexible than that of large ones; thus they can be located more easily, strategically close to points of demand. This should reduce transmission costs and losses, and additionally may allow inputs to be injected into an electricity supply network, at locations where they are most needed to balance the system and to compensate for losses.

Small schemes take far less time than large ones to plan, finance, design and build. Hence the investment can be more easily tailored to suit energy demand and to suit the inevitable changing and unreliable forecasts of demand. When a large scheme is commissioned, it causes a large step-up or steps-up in the capacity of the system. Even if a large scheme is commissioned at the optimum time it can be inferred that, previously, there would have been a power and/or energy shortage or that just after commissioning there would be a power and/or energy surplus (or both). In any event, for a time the large scheme is not fully investment-effective, while the commissioning of small schemes can be fitted more closely to the demand curve. Since the run-up period to the commission of small schemes is only one third or so of that for large schemes (say, two to three years compared with six to ten) they can be added to the system or cancelled, to suit changing demand, at much shorter notice. Their introduction into a programme of development makes planning to meet a range of possible demand forecasts more feasible and flexible. The point is illustrated in Fig. 16.2. For both small and large hydro projects, however, the importance of regular and reliable updating of demand forecasts ought to be emphasized, and this is apparent from the figure.

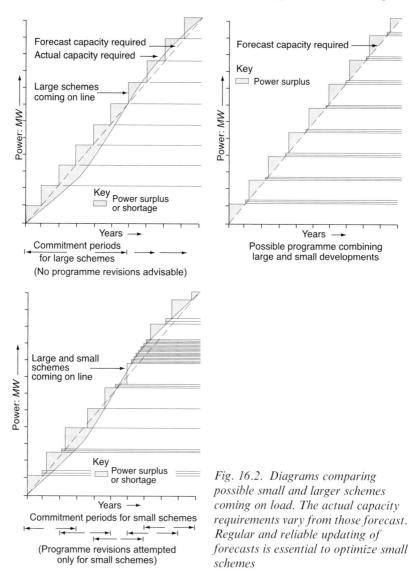

Fig. 16.2. Diagrams comparing possible small and larger schemes coming on load. The actual capacity requirements vary from those forecast. Regular and reliable updating of forecasts is essential to optimize small schemes

Under favourable circumstances, small schemes can be planned and implemented and can pay back their capital investment during the years before being eventually superseded by larger schemes. Thus there is a need to be able to add relatively small increments to a system for supply of electricity at short notice, in order to tailor it to demand.

Where no competitive alternative is available
In suitable circumstances, a small hydropower facility may be the most economic source of power and energy to meet isolated demands, such as from remote villages and factories. This may be the case where the national grid is not sufficiently close for economic connection. Where there is a potential for developing small (or large) hydro, it has the previously mentioned advantages over most other means of energy generation, i.e. of being an indigenous renewable source. Thus, both in developed and developing countries, there is a need for small hydropower to meet demands remote from a main supply line.

Provision of relatively small amounts of power and energy at modest capital cost
This heading speaks for itself in situations where the unavailability of finance restricts the opportunity to develop larger power projects. It should also be borne in mind that the recurrent fuel costs for thermal power plants also require financing before there has been sufficient revenue from the plants to repay the other costs involved.

Different users
There are three main categories of users of small hydropower: national electricity authorities; private enterprises; and remote towns and villages. Each is now covered in more detail.

National electricity authorities
National electricity authorities adopt attitudes to the use of small hydro according to their individual circumstances. For example, in China, small hydro is relied on to meet a substantial part of the national power demand. Considerable problems must be encountered in meeting peak demands, in co-ordinating the supplies from so many small run-of-river contributions, as well as in the provision of stand-by plant, to cover plant outages in so many small stations. Independent county grids have been emerging, perhaps integrated into the state grid through one or two points, perhaps as a response to these problems. Defects in the past have been poor performance, poor

quality of electricity and high maintenance costs. There has been a move towards more efficient automation.

In Malaysia, Indonesia and the Philippines, for examples, small hydro is seen as particularly attractive by the national electricity authorities, as a way of supplying towns and villages beyond the current reach of the grid. Aid, and finance on concessionary terms, have been provided for this by several western countries.

In Sri Lanka, the Ceylon Electricity Board (CEB) has been wary of becoming involved in small hydro development, while it still has larger and more manageable hydro schemes to promote. It feared that small hydro supply would be of a run-of-river nature, inadequate to meet full demands, and so would leave CEB to supply the awkward-to-manage, unfulfilled irregular balance of requirements. In some countries, such as India and the United States, small hydro is largely seen as most suitable as an adjunct and subsidiary to schemes for irrigation, water supply and canal and river waterways.

In the UK, the Government introduced the Non-Fossil Fuel Obligation (NFFO) in 1990, which provides for the purchase of energy supplied to the grid at an advantageous rate linked to the retail price index for a fixed period of years. However, the cost and delay involved in getting contracts approved, seemed to be beyond the resources of most small producers, as elaborated by Professor E. M. Wilson [5].

The North of Scotland Hydro-Electric Board (becoming Scottish Hydro-Electric, then Scottish Power plc, then Scottish & Southern Energy plc — at least until 2001!) would have liked to promote small hydro schemes with outputs not less than 500 kW, in conjunction with various industries (e.g. fish farms, saw mills, distilleries, hotels). This is considered desirable in order to regulate the load, facilitate economic operation of its system and minimize the cost of day-to-day operation and promote simple maintenance. The electricity system in Scotland had more than sufficient plant, so additional storage in small schemes was superfluous in the short term, except for 2 to 4% storage to impound minor spates (high flows), which was economically a very attractive provision. The Board would nevertheless have liked to construct run-of-river schemes to save fossil fuel, but was prevented from constructing such new schemes by government restrictions on spending. One scheme that was promoted was deferred for financial reasons. Apart from that difficulty, the promoters were faced with numerous 'objections', mainly on environmental grounds. Therefore the way ahead in this field is not easy for the electricity-generating companies in the UK, although they are required to purchase electricity from private generators.

Perhaps enough has been said here to illustrate the diversity of uses and approaches among national electricity authorities. Everywhere, attitudes and initiatives are no doubt influenced by political thinking and economics. Regrettably, economics has so many variables and differing schools that economists are able to adapt the rules to get the answers that they or their employers prefer. Perhaps we should be seeking an alternative standard, a qualitative scale on which competing projects can be awarded marks for factors such as the value of annual production, number of beneficiaries, annual cost, safety, reliability, positive or negative effects on the environment, etc. It should then be possible for decision makers to judge whether the total, and the balance of marks, fulfil their objectives. With this in mind, the Author devised a system for the classification and selection of small hydro projects, which was used for the reconnaissance, pre-feasibility and feasibility studies of 15 sites on six different Indonesian islands. The system will be described presently, but the point to be made here is that, despite providing measurements of numerous characteristics of the projects, it proved to be the economic and financial characteristics that mainly influenced the authorities and financing agencies in their choice of the schemes that were to go ahead.

Private enterprises

While in some countries, such as China, the development of small hydropower by private users has been actively encouraged by the government, in others it has been impeded by regulations, licensing fees, the need for government approvals and lack of encouragement to supply surplus energy to the grid. Such problems can overwhelm the attractions of development for a small user. In one river basin in the USA, potential developers got together in a co-operative effort to overcome such problems and to co-ordinate development. Unfortunately, divisive influences and lack of government encouragement caused this promising collaboration to dissolve.

What is needed, but is often lacking, is government support in providing advice and assistance, including financial inducements, because of the benefits mentioned earlier of developing a national asset. Of course the cost to government has to be fairly weighed against the finance available, and the other uses for which the finance could be applied.

Clearly there are often sufficient net benefits to justify private users to develop small hydropower, if the potential users are prepared to take the necessary time and trouble. In Sri Lanka, there are several hundred schemes of 15 to 250 kW, implemented in the past by private enterprise for tea estates. Most of these schemes eventually fell into

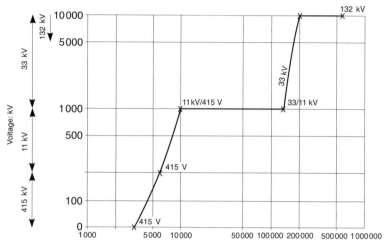

Fig. 16.3. Possible costs and voltages for connecting small hydro to a transmission grid estimated for Scotland in 1985 (see text for further details)

disrepair and have come under national ownership. A programme to renovate many with international and bilateral aid was put in hand with the prospect of reasonably low cost and rapid pay back. Regrettably this became subject to delays and the Author is not aware of how it now stands.

Connection to a grid system can be of great advantage to small hydro owners, either as an outlet for surplus supply, or as a source of power and energy to meet deficiencies due to outages of the small hydro. As an indication, Fig. 16.3 gives costs of connections to the transmission mains estimated for Scotland in 1985. Subject to changes in cost indices since then, it would probably be reasonable to add some 66% to the figures indicated, to convert them to year 2000 costs. In addition, the cost of overhead lines has to be added, estimated for 1985 at approximately £8000 per km for an 11 kV circuit, £12 000 per km for a 33 kV circuit and £60 000 per km for a 132 kV circuit tower line. Metering costs then added about £500.

In the UK, there is a National Association of Water Power Users, chaired by E. M. Wilson.

Remote towns and villages

Electricity supply to remote towns and villages is clearly a national responsibility and is usually accepted as such. However, a source of supply could be developed by local initiative, either by co-operative effort or by the initiative of the town or district council. The Author is not, however, familiar with any such instances.

Generally, where feasible, new small hydro is found to be competitive with diesel, e.g. in 1985 in Scotland, projected small hydro ranged in cost from 1p to 5p per unit of energy (2·5p or below being considered economic) compared with 4p to 8p per unit for diesel. Transport and storage of diesel fuel can be important considerations in the comparison.

Classifying and selecting small hydro schemes

The system of classification mentioned above was intended to be comprehensive, flexible and practical for use at reconnaissance, prefeasibility or feasibility levels of evaluation. It could easily be adapted and applied in the evaluation of small, medium or large hydropower projects, and is suitable for processing and presentation by computer.

The 15 sites on six different Indonesian islands to which the system was applied, were generally widely separated. Access to some was difficult and few had significant characteristics (such as values of available discharge and potential head) in common. At all but two sites, the aim was to replace or supplement existing diesel generators in remote load centres, with hydropower generators. At the remaining two sites, consideration was given to installing hydro plant at existing dams already connected to the local grid. The installed capacities under consideration were up to about 2·5 MW.

Sites were identified from map studies, existing reports and/or surveys. A reconnaissance study of each site was carried out as soon as possible by a small multi-disciplinary team comprising a senior civil engineer, a hydrologist and a geologist. To avoid abortive work and expense, sites found to be unsuitable for reasons of excessive cost or serious geological difficulties, were eliminated as quickly as possible.

One of the objectives of the initial study was to formulate the classification system to be used. The system devised is set out in the table shown in Fig. 16.4. Thirty characteristics relevant to classification are listed in the second column. The measurement of some of these characteristics is inherent in the measurement of others. Of the 30, only 14 or fewer need to be considered in the selection of a site. Nine characteristics asterisked for use in selection are the following:

1   Potential demand
16 Scope for standardization
19 Foreign cost, present value
20 Local cost, present value
24 Net present value at 10% discount rate
25 Benefit/cost ratio at 10% discount rate

1. Ref. No.	2. Characteristic	3. How measured	4. Value	5. Classification A	5. Classification B	5. Classification C	6. * If useful for selection
1.	Potential demand	Forecast for 1995 (kW)		>1000	500–1000	<500	*
2.	Gross head	Accounted for by resulting energy output under item 7 (m)		>50	10–50	1–10	
3.	Storage	As above (Type)		Seasonal	Daily	Run of river	
4.	Access	Length of new road or major upgrading required (km) included in cost under items 15–20 & 24–28 and time, item 30		<2	2–5	>5	
5.	Transmission type	Station		Grid connection	Isolated load	–	
6.	Transmission length	Length of connection to existing line (km)		<10	10–20	>20	
7.	Potential energy output	Minimum energy available from the proposed installation (GWh/year)		>7·5	3–7·5	<3	
8.	Absorbable energy	Energy absorbable by 1995 forecast demand (GWh/year)		>3	1–3	<3	
9.	Installed power	Maximum economic installed capacity (kW)		>1000	500–1000	<500	
10.	Firm capacity	Maximum daily peak demand forecast for 1995 that can be met with 90 per cent probability (kW)		>1000	250–1000	<250	
11.	Compatability of potential output with potential demand	Annual energy output as percentage of forecast demand in 1995 (per cent)		90–120	120–150 or 60–90	>150 or <60	

Small hydropower

		Not available	Partly	Existing	
12.	Alternative means of satisfying demand				
13.	Sufficiency of data	Most necessary data available	Missing data rapidly obtainable	Vital data unobtainable in reasonable time	
14.	Geotechnical geological difficulties	Geologically suitable	More investigations needed	Geologically unsuitable	
15.	Environmental impact	Minor	Modest	Major	
16.	Scope for standardization Resulting saving in cost in items 19 & 20 which seems possible				
P	(per cent foreign cost)	>10	5–10	<5	* *
Q	(per cent local cost)	>10	5–10	<5	
17.	Scope for grouping As above				
P	(per cent foreign cost)	>10	5–10	<5	
Q	(per cent local cost)	>10	5–10	<5	
18.	Scope for local manufacture Proportion of total E&M cost which could be local manufacture (per cent)	80–100	20–80	<20	
19.	Foreign cost present value Capital cost, including engineering and contingencies (£1000)	<100	100–500	>500	*
20.	Local unit, present value Capital cost including land, compensation and contingencies (£ × 10)	<500	50–2000	>2000	*
21.	Cost of energy at 10 per cent discount rate Cost per kWh of absorbable energy forecast for 1995 (£/kWh)	<100	100–300	>200	

Fig. 16.4. Classification and selection system

1. Ref. No.	2. Characteristic	3. How measured	4. Value	5. Classification			6. * If useful for selection
				A	B	C	
22.	Cost per kW of installed capacity	Capital cost divided by installed capacity (£/kW)		<1000	1000–2000	>2000	
23.	Cost per kW of firm capacity	Capital cost divided by firm capacity (£/kW)		<2500	2500–5000	>5000	
24.	Net present value at 10 per cent discount rate costs (£1000)	Difference between present of lifetime benefits and present values of (£1000)	>2000	0–2000	<0		*
25.	Benefit/cost ratio at 10 per cent discount rate	Cost of equivalent diesel or grid energy and capacity divided by cost of hydro (ratio)		>1·5	1·0–1·5	<1	*
26.	Financial return on investment	Financial internal rate of return (per cent)		>12	8–12	<8	*
27.	Economic return on investment	Economic internal rate of return (per cent)		>12	8–12	<8	*
28.	Pay-back period at 10 per cent discount	(Year)		<5	5–10	>10	
29.	Special factors	Client's assessment		Favourable	–	Unfavourable	*
30.	Commissioning time	Time required to investigate and implement (months)		<25	25–36	>36	

26 Financial return on investment
27 Economic return on investment
29 Special factors

Six of the above are measures of cost and economic or financial return, but many of the remaining 21 are measures representing the characteristics of the sites. To appraise the relative merits of a site, the most promising scheme foreseeable at that site has to be identified in outline and then evaluated. The most promising scheme is taken to be that with the highest net present value. The values of each characteristic are then tabulated under a classification of A, B or C to which a range of values for each class has been assigned, with the intention that schemes with a classification 'A' are likely to be more economically or otherwise attractive and schemes with classifications B and C progressively less so. This does not mean, however, that schemes with one or more B or C rankings would not be selected, if other factors were particularly favourable. In some characteristics, Class A might not be wanted, for example under the characteristic 'installed power', if an objective were to implement only very small schemes.

The system of classification is designed to form a computerized database, but for a small number of potential schemes it is suitable for keeping in dossier form. The system makes it possible to identify and compare all schemes fulfilling whatever criteria it is required to meet. For example, if it were desired to find a scheme with more than 1000 kW installed capacity, with a high economic return on investment, all schemes with classification 9A and 27A can be called up and considered. If no scheme were ranked 27A, those with rank 27B could be considered. The classification of the schemes thus selected could also be called up and taken into consideration when required under items 1, 8, 9, 11, 16 (P&Q), 18–20, 24–27 and 29.

To make a full appraisal of schemes of interest, other or all items may be consulted, some helping to explain the values of others. For example, long access, transmission lines or geological difficulties would help to explain high cost.

Conclusions

(a) Small hydropower can be developed successfully and competitively in its own right. A principal requirement is that projects must be soundly planned and implemented. The following may be among the desirable objectives:
   (i) multiple developments in series or parallel
   (ii) connections to grid systems
   (iii) conjunctive operation with other energy sources

(iv) shared facilities by incorporation in other-purpose projects
(v) automation
(vi) reliable, trouble-free operation.

To paraphrase a comment in a World Bank publication [6], lower initial costs do not necessarily mean lower long-term costs if they are obtained at the expense of increased maintenance, lower reliability and shortened project life.

(b) Sufficient provisions must be made for the maintenance of small schemes, by training and organization.
(c) There are basic needs to justify the development of small hydro:
   (i) as a means of capitalizing a natural resource and national asset
   (ii) to complement energy and power from larger sources, so providing greater flexibility and accuracy in matching demand
   (iii) to meet discrete demands for power and energy, where small hydro is the best method of supply.
(d) Small hydropower can be a facility of economic benefit in meeting the needs of public electricity supply authorities, private enterprise and remote towns and villages, in other words, of a wide range of possible beneficiaries. While small hydro must be prepared to compete, on equal terms, with alternative means of supply, national governments should desirably sponsor pilot or demonstration schemes and, with conviction on the intrinsic worth of small hydro, should be prepared to provide administrative, technical and financial assistance to help and encourage those potential developers who have limited resources.

Further information

In *International Water Power & Dam Construction*, November 1999, it was reported that ten member countries of the International Energy Agency (IEA), including the UK, have set up a small hydro atlas to provide a global database of sites with emphasis on projects $>50\,kW$ and $<10\,MW$ with the objective of enhancing communication and promoting developments in the industry. Visit: www.small-hydro.com

References

[1] World Bank. 'A methodology for regional assessment of small scale hydropower' (Energy Department Paper No. 14, May 1984).
[2] World Bank. 'Small hydroelectric components in irrigation and water supply projects' (Energy Department Note No. 60, July 1985).
[3] World Bank 1980.
[4] International Bank of Reconstruction and Development, 1980 (ibid of [3])
[5] Wilson, E. M. Problems and risks of smal hydro development in the UK. *International Journal of Hydropower and Dams*, September 1994, Vol. 1, pp. 17–18.
[6] International Bank of Reconstruction and Development Publication.

17. Safety and inspection of reservoirs

Introduction

The safety of reservoirs in the UK was first covered by the Reservoirs (Safety Provisions) Act of 1930 [1], which was drafted following two catastrophic reservoir failures in 1925, at Skelmorlie (5 m high) in Scotland and Coedty (11 m high) in Wales. The Act prohibited the design and supervision of construction of 'large reservoirs' in Great Britain, except under the supervision of a qualified civil engineer, defined in the Act as a civil engineer appointed by the Secretary of State, to be a member of a panel of engineers whose competence and experience qualify them to undertake the duties placed on them in the Act. A 'large reservoir' was defined under the Act as one capable of containing 5 000 000 gallons of water above the lowest level of the surrounding ground. While 26 lives were lost due to dam failures in the UK in the twentieth century before 1926, and 402 lives were lost between 1799 and 1900, none have been lost since 1925. Three failures with the greatest number of lives lost were those at Coedty in 1925 (16 lives lost), Dale Dyke in 1864 (245 lives lost) and Bilberry in 1852 (81 lives lost).

In 1933, the Institution of Civil Engineers published an *Interim Report of the Committee on Floods in Relation to Reservoir Practice* [2], which was intended to advise on the maximum intensity of flood which should be provided for. This report gave an empirical enveloping curve of flood intensities related to catchment areas, for upland areas. With some additional flood records, including those of the 1952 Lynmouth disaster, it was published again in 1960 [2]. Until the mid-1970s, this was the basis dam engineers used to assess both design flood and freeboard for UK dams. In 1965, a Committee on Floods in the UK was set up by the Institution to review the whole question of floods. As a result of a recommendation by this Committee, a national floods study was undertaken by the Natural Environment Research council

(NERC). The studies, which were extended to the Republic of Ireland, were published in five volumes by NERC in 1975 [3]. In 1978, the Institution of Civil Engineers published *Floods and Reservoir Safety: An Engineering Guide* [4] which was intended 'to assist those individuals who bear the personal responsibility that comes from being appointed to the statutory panel of engineers qualified to design and inspect reservoirs'. The engineering guide was described briefly in Chapter 6 on the subject of hydrological studies and will be returned to presently.

It defines standards to cover all major points of principle. The recommendations made are not in any way mandatory. Nevertheless, they have been universally followed as guidelines by engineers for reservoirs in the UK.

The Reservoirs Act 1975 [5] repealed the Reservoirs (Safety Provisions) Act 1930, but re-enacted the legislation in a strengthened and more effective form. It was not, however, brought into full effect, in stages, by statutory regulations, until January 1987. In addition to the quite comprehensive provisions in the 1930 Act, the 1975 Act provides for the first time:

(a) a duty of enforcement by local authorities, and general powers of supervision by the Secretary of State
(b) for large raised reservoirs to be kept under continual surveillance by a qualified engineer
(c) the steps to be taken to remove a reservoir from the scope of the Act, or to secure the safe abandonment of use of a reservoir.

The powers and duties of three distinct organizations or persons: the undertaker, the enforcement authority and the qualified civil engineer, are described in order to ensure the safe and efficient design, construction and use of reservoirs. A 'panel' system of qualified civil engineers is the linchpin of reservoir safety, including certification and periodical inspection. 'A large raised reservoir' under the 1975 Act became one 'impounding a volume of 25 000 cubic metres or more' above the natural level of any part of the land adjoining the reservoir. Lagoons and tips for receiving refuse in solution or suspension are the province of the Mines and Quarries (Tips) Act 1969 and are excluded from the 1975 Act.

It is believed that the Reservoirs Act applies to about 2450 large raised reservoirs in the UK, of which about 80% (1960) are embankment dams and 20% (490) are concrete or masonry. Nearly 50% of the embankment dams are 5 m or less in height and only 20% are greater than 15 m high. Some 70% were built before the start of the twentieth century. Some 36 dams are recorded as having failed from 1798 to 1900 and 24 from 1900 to 1971 (there were ten catastrophic failures 1960–1971, but without loss of life). Of the failures of dams, most were under 10 m high and only four were significantly more than 20 m high: these were Blackbrook and Slaithwaite in 1798, Dale Dyke, 29 m high in 1864 and Carsington (during construction) in 1984.

In 1990, the Building Research Establishment (BRE) published *An Engineering Guide to the Safety of Embankment Dams in the United Kingdom* [6]. In the Author's opinion, this guide is pointed more to the guidance of supervising engineers responsible for surveillance of reservoirs than to guidance of qualified engineers responsible for design, supervision of construction, statutory inspections and certification of reservoirs, required under the Act.

In 1991, BRE published *An Engineering Guide to Seismic Risk to Dams in the United Kingdom* [7], which was referred to in some detail in Chapter 9. CIRIA has also published an engineering guide to the safety of concrete and masonry dam structures [8]. The guide is primarily addressed to members of the panel of qualified supervising engineers.

So, as we have seen, high standards are applied to ensuring the safety of dams in the UK. This is because the existence of dams does create hazards to lives and properties and those lives and properties have to be protected against the risks of failure. The consequences of failure of a nuclear installation would be more far-reaching and prolonged than those of failure of a dam which, nevertheless, could have disastrous consequences, cause a public outcry and result in prosecution of any one responsible if believed to have been negligent.

The keys to the safety of reservoirs are that they should have been soundly designed, their construction and operation should have been consistent with the design and they should be subject to regular inspection by a qualified engineer. Some existing reservoirs were built long ago, and records of their design and construction are lacking. It has been necessary for Inspecting Engineers under the Reservoirs Act to satisfy themselves that dams under the Act are safe, and where data sufficient to do so are lacking, to obtain new data; where necessary, this might involve site investigations, surveys and analysis. Inspectors have to be on the look out for signs of potential failure, notably from indications from instrumental monitoring or signs of increasing leakage and/or settlement.

The Reservoirs Act 1975

The Reservoirs Act 1975 is the principal instrument governing the provisions for safety of dams in the UK and its introduction has been described here. A guide to the Reservoirs Act 1975 has been published [9] with the purpose of providing guidance on the application of the Act, reflecting current views and practice of the dam engineering profession. The contents of the Act will now be summarized.

The Greater London Council (now Greater London Authority) and county councils in England and Wales, and the regional and island councils in Scotland, are required by the Act to maintain a register of large raised reservoirs. These councils are also enforcement authorities, except where they themselves are also 'undertakers'. In this case, if the reservoir lies wholly within the area of the local authority which is its undertaker, no enforcement authority is required. Local authorities are required, both as enforcement authorities and undertakers (namely, owners), to report to the Secretary of State at prescribed intervals, on the steps which they have taken to ensure that the Act is complied with. The Secretary of State has the power, if satisfied that the local authority is in default, to make an order, directing it to carry out its duties in accordance with the Act.

A qualified civil engineer is defined (for the purposes of the Act), as a member of one or more panels of civil engineers set up under the Act by the Secretary of State, after consultation with the President of the Institution of Civil Engineers. Qualification in the first instance involves passing an interview, with two or three eminent engineers appointed by the Reservoirs Committee of the Institution. Appointments to panels run for five years and members have to apply and satisfy the qualification and fitness requirements, if they wish for re-appointment for a further term. The existing panels are:

- The All Reservoirs Panel (Panel AR), for the inspection and certification of reservoirs of all types and sizes within the Act
- The Non-impounding Reservoirs Panel (Panel NIR)
- The Service Reservoirs Panel (Panel SR)
- The Supervising Engineers Panel (Panel SupE)

Under the Act, the qualified civil engineer has broadly four statutory labels: 'the Construction Engineer', 'the Inspecting Engineer', 'the Supervising Engineer' and simply, 'the qualified civil engineer'. The panel system ensures that, for each function, the engineer has appropriate qualifications and experience.

The Construction Engineer

The 'Construction Engineer' is the qualified civil engineer employed by the undertakers:

(a) to design and supervise the construction of a large raised reservoir, whether it is new construction or alterations which increase capacity
(b) to give preliminary, interim and final certificates
(c) to report on and supervise the re-use of an abandoned reservoir.

The Construction Engineer has responsibility for issuing preliminary certificates, allowing the initial filling and operating conditions of a large raised reservoir, and a final certificate, allowing and governing the final filling and operating conditions. The preliminary certificate(s) may be expected to propose interim level(s) to which the reservoir may be filled, and restrictions on the rate of filling. The final certificate would normally be expected to allow filling to the maximum water levels for which the reservoir was designed under the designed operating conditions. The Act allows successive preliminary certificates to be issued, but prohibits the issue of a final certificate earlier than three years from the issue of the first preliminary certificate. This time interval is intended to allow a sufficient period of observation during the crucial early years of a reservoir. The Construction Engineer has a duty to explain to the undertakers why a final certificate has not yet been issued when five years have elapsed since the issue of the first preliminary certificate, and a copy of the explanation must go to the enforcement authority. The Construction Engineer is required to annex to the final certificate a note of any matters which it is considered should be watched by the Supervising Engineer, until the first periodical inspection of the reservoir is made.

The Act introduced an 'interim certificate', for use by the Construction Engineer engaged on works, to increase the capacity of an existing large raised reservoir, so that, pending the issue of a preliminary certificate, the statutory filling level and operating conditions of the existing reservoir might be revised.

The Construction Engineer and the Inspecting Engineer are both empowered to specify the frequency and manner in which the prescribed record of a reservoir is to be given.

If the Construction Engineer is dealing with a reservoir that has been completed and filled with water for at least three years, and is satisfied that it is sound and satisfactory and may be safely used for the storage of water, a final certificate may be issued forthwith.

The Inspecting Engineer

Once commissioning is completed and the Construction Engineer has issued the final certificate, the reservoir is subject to a periodical inspection procedure, and the qualified civil engineer, in the role of 'Inspecting Engineer', is responsible for recommending measures to ensure that it is operated and maintained in a safe manner.

The Inspecting Engineer must not be an employee of the undertaker, must not be the engineer who acted as the Construction Engineer and, when acting as the Inspecting Engineer, must not have any working or business connection with the Construction Engineer. (Guidance on these constraints requiring 'independence' of the inspecting engineer are given in Part B 10(1) and 10(9) [10].)

Two certificates have to be issued following a periodical inspection. In the first, the Inspecting Engineer certifies whether the report does or does not include recommendations as to measures to be taken in the interests of safety. If it does, in the second, the qualified civil engineer, who must be appointed to supervise the implementation of the measures, and who may be an employee of the undertaker, certifies that the measures have been carried into effect.

The Inspecting Engineer's report would pick up those matters noted by the Construction Engineer or a previous Inspecting Engineer, and revise, amend or add to them, thereby ensuring that instructions to the Supervising Engineer are continually updated, to take account of those changes liable to occur in the reservoir structure, the catchment, and the area downstream with the passage of time.

The Inspecting Engineer has a duty to inspect and report on the results of an inspection, including recommendations as to the time of the next inspection (not later than after an interval of 10 years) and as to recommendations made in the interests of safety (which thereby become mandatory). The subject of monitoring and inspection and the possible associated instrumentation will be returned to in Chapter 18.

The Supervising Engineer

Once constructed, the reservoir must be kept under continual surveillance by a Supervising Engineer who is not only capable of interpreting operating data and records, but also has a trained eye to notice and assess the effects of an unexpected event on reservoir safety. Once the Construction Engineer has issued the final certificate, a duty is placed on undertakers to appoint a SupE Panel Engineer, to be responsible for the day-to-day oversight of matters affecting the safe operation and behaviour of the reservoir. In this SupE Panel Engineers are guided initially by the matters drawn to their attention in the Construction Engineer's note, and subsequently in the Inspecting Engineer's periodical reports. They have to advise the undertakers of any aspect of the reservoir's behaviour affecting safety, to draw their attention to breaches of the Act (particularly as to any breaches of compliance with requirements related to the storage of water), and to call for a periodical inspection at any time thought fit. At least once a year, they must give the undertakers a written account of the action they have taken to implement the instructions of the Construction and the Inspecting Engineers. In practice, the number of times that they visit the reservoir in the course of a year will depend on the condition of the reservoir and on any work on it which is in progress.

It is a matter for them in agreement with the undertakers, and may be influenced by the recommendations of the Construction and/or Inspecting Engineer(s).

Copies of the SupE accounts of their work are not normally required to be sent to the enforcement authority. Where, however, the SupE advises the undertakers to arrange for an inspection by the Inspecting Engineer under section 10, or to take any action in connection with recommended safety measures, copies of that advice must be sent, by the SupE, to the enforcement authority. Similar action is also required where the SupE draws the attention of the undertakers to a breach of the mandatory conditions for the storage of water in the reservoir, or of the statutory requirement related to the keeping of records of prescribed information.

Qualified civil engineer (in general)

Apart from the defined roles as Construction, Inspecting and/or Supervising Engineer, only a qualified civil engineer may:

(a) be employed by the enforcement authority, when it uses reserve powers in default of an undertaker, or where the authority takes emergency action
(b) approve and certify the satisfactory completion of alterations, to remove a large raised reservoir from the ambit of the Act
(c) report on measures to be taken to ensure that a reservoir, when its use is abandoned, is incapable of filling accidentally or naturally, so as to constitute a risk to public safety
(d) advise the enforcement authority on the time to be specified, in a default notice served on the undertakers, requiring them (1) to carry out works or measures before a reservoir is brought back into use, or (2) to implement any measures recommended by an inspecting engineer
(e) act as an independent referee on disputes.

The qualified civil engineer, whatever function is being performed (and not the undertakers), is required to send all certificates and annexes, as well as copies of reports (except a report of an Inspecting Engineer which makes no recommendation as to safety measures) to the enforcement authority.

An enforcement authority may consult any civil engineer with regard to the time in which measures shall be carried into effect to secure the safe abandonment of use of a reservoir, so long as they do not involve any alterations. Similarly, any civil engineer may design the alteration required to remove a reservoir from the ambit of the act, but the alterations must be approved and supervised by a qualified civil engineer. Although there is no requirement for the undertakers to employ a qualified civil engineer to design and supervise alterations which do not increase the capacity of a large raised reservoir, they are required, after the alterations have been completed, if they have not been designed and supervised by a qualified engineer, to have the reservoir inspected, if such alterations might affect its safety.

Undertakers

Ultimate responsibility for the safety of a reservoir rests with the undertakers.

Any civil engineer may design the alterations required to remove a reservoir from the ambit of the Act, but the alterations must be approved and supervised by a qualified civil engineer. Where the abandonment of use of a reservoir does not involve alterations, the reservoir remains on the register and is subject to the inspection and supervision requirements of the Act. There is also no requirement for the undertakers to employ a qualified civil engineer to design and supervise alterations which do not increase the capacity of a large raised reservoir, but after the alterations have been done, if they have not been designed and supervised by a qualified civil engineer, if the alterations might affect its safety, the reservoir must be inspected by an Inspecting Engineer.

Undertakers are required to arrange for a first periodical inspection of a new reservoir within two years of the issue of the final certificate, thereby ensuring that a reservoir is inspected by an independent qualified civil engineer in the critical early years of its life. The undertakers are placed under an obligation to implement, as soon as practicable, any measures recommended in the interests of safety by an Inspecting Engineer. The qualified civil engineer employed to supervise the implementation of the measures must issue and supply them with a copy of a certificate, confirming that the measures have been implemented. The undertakers have a right to refer to a referee complaints about recommendations made by an Inspecting Engineer in any case, or by a qualified civil engineer who has been appointed as a result of enforcement action, or has reported on either the re-use or on the abandonment of use. The referee is required to issue a certificate, with a copy to the enforcement authority, stating whether the referee's decision does or does not amend an engineer's report.

A duty is placed on undertakers to provide any necessary instrumentation required for the proper recording of essential information about the operation and behaviour of a reservoir, and to comply with any directions made by the Construction Engineer or the Inspecting Engineer, as to the frequency and manner in which the information is given.

Failure to comply with the Act constitutes a criminal offence. The more serious failures are punishable on indictment by unlimited fines. Fines on summary conviction originally ranged from £100 to £800. The Author is not aware of any change in these. The fine of £800 relates to knowingly or recklessly giving, or making use of, false information to any qualified civil engineer employed for different purposes under the Act. The liability for an offence applies to officers and members of a corporate body as well as to that body itself. Proceedings under the Act may be instituted by any local authority in whose area the reservoir, or part of it, is situated, and by the Secretary of State, but otherwise not without the consent of the Director of Public Prosecutions.

Enforcement authorities

While technical matters are regarded as the preserve of the qualified civil engineer, enforcement is regarded as a legal and administrative matter to be dealt with primarily at local level. Local authorities have a duty placed on them to see that undertakers observe and comply with the Act. A register of all large raised reservoirs in their areas is to be kept by local authorities, and these registers should by now have been completed. A considerable amount of information has to be provided to either local or enforcement authorities, including:

(a) notices of intention to construct a large raised reservoir, or to bring one back into use, or in the case of use being abandoned
(b) details of appointment of Inspecting Engineers; and appointment, and cessation of appointment, of Supervising Engineers
(c) copies of all certificates and reports issued by qualified civil engineers
(d) copies of explanations as to why a final certificate has not been issued; and copies of advice given by the Supervising Engineer.

Information to be included in the registers is proscribed by regulations. The registers are available for inspection by any member of the public, who may, with the consent of the Director of Public Prosecutions, institute a prosecution. It was intended that information on the registers should be limited to that necessary to enable an interested person to ascertain whether the Act is being complied with, and would not include details of design or layout. In the event of non-compliance with the Act, enforcement authorities are empowered to require undertakers to appoint a qualified civil engineer:

(a) to act as construction engineer for reservoirs in course of construction
(b) to inspect, report on and supervise an abandoned reservoir that has been brought back into use

(c) to carry out any periodical inspection as required by the Act, or to implement measures recommended by an Inspecting Engineer in the interests of safety
(d) to act as Supervising Engineer
(e) to report on the measures to be taken to ensure the safe abandonment of use of a large reservoir.

If the undertakers fail to appoint a qualified civil engineer as required, then the enforcement authorities may undertake the appointments themselves.

The enforcement authority may also take emergency action to protect life and property when a large raised reservoir appears to be dangerous, but only in accordance with the recommendations and under the supervision of a qualified civil engineer. Reasonable costs incurred by enforcement authorities in exercising their reserve powers, in the event of default by an undertaker, and in taking emergency action, are charged on the undertakers.

In order that enforcement authorities may carry out their functions under the Act, enabling powers are provided to allow an authority's authorized agent entry onto land, subject to the giving of seven days notice, for the purpose of:

(a) finding out whether a reservoir is a large raised reservoir, or is one under construction, or is in use as a reservoir
(b) taking enforcement action
(c) using their reserve powers to undertake appointments under Section 15, as just noted (under (a) to (e))
(d) taking emergency action, in which case there is no requirement to give seven days notice prior to entry, including onto neighbouring land.

Although enforcement authorities are given extensive powers, no enforcement provision breaches the principle that technical matters are the sole responsibility of the qualified civil engineer. It is therefore no impediment to effective enforcement that many local authorities and their officers may have had little previous experience of dealing with reservoirs.

Interpretation of the Act, if required, rests with the Courts but the Author is not aware of any respects in which it has so far been tested.

It will probably be no surprise that some owners of lakes were quite unaware that their lakes were large raised reservoirs, and that they were accordingly to be faced with quite heavy responsibilities under the Act. The Department of the Environment has in fact asked Inspecting Engineers to interpret the Act leniently in favour of county land owners who are owners of large raised reservoirs that are impounded by dams of modest height and which pose no likely threat to persons or property.

Proposed changes

Some changes to the Act and/or related regulations have been considered over a lengthy period but, in the main, have been dropped. A review of the Reservoirs Act has been put in hand by the Reservoirs Committee at the Institution of Civil Engineers; the following changes are expected.

- The powers of enforcement authorities are likely to be transferred to the independent Environment Agency which has taken over the functions of the National Rivers Authority (NRA).
- It is proposed to remove from requirements of the Act, silted up reservoirs holding less than 25 000 cubic metres, which are certified by a qualified civil engineer to present no greater risk than if they had originally been designed to hold their present capacity.
- Reservoirs holding 100 000 cubic metres or less, which are certified by a qualified civil engineer to pose a very low hazard to people and property, are expected to become exempt from the requirement of continuous supervision.

Reviews by panels of experts

In order to obtain a review and a second opinion on the soundness of design and construction of new reservoirs, it has become recognized practice for the undertaker, and/or financing agency, to appoint an independent panel of experts to visit the design office and construction sites at key times (and usually at intervals not longer than six months) and to report on progress and safety issues. Usually the leader of the panel is a civil engineer with the relevant experience and, to provide continuity and co-ordination, that appointment should continue throughout design and construction. Other members of the panel are usually specialists who may change according to the relevance of their discipline as work progresses. Usually they do not number more than three, and may include at times, according to need, a geologist, hydrologist, economist, construction expert, plant expert, and so on. The panel usually does not have any executive responsibility but, if the members are well chosen, their advice can be valuable to the undertaker, to the engineer and to the contractor; it is unlikely to be ignored, unless for very good reason.

Technical Guidelines for the appointment of panels of experts, and other provisions for the review of the design, construction, maintenance and safety of dams, were published in September 1983 [11] by the Agricultural and Rural Development Department of the World Bank (though not to be quoted as representing the views of the World Bank). The Guidelines envisaged the appointment of panels comprising two to four experts, including a dam generalist and an engineering geologist. A specialist in the type of dam selected might be included as would often a hydraulic engineering expert. The panel would be supported by consultant experts on special problems. The Guidelines expect the panel for review of design to be independent of the designers, except for one member of the consulting organization appointed by the owner.

Remedial works
Recommendations that might result from inspections or review include the following.

(a) Increase in flood discharge capacity. In order not to interfere with the operation of the existing spillway, this is usually best achieved by introducing a separate new spillway, which might be constructed in the rim of the reservoir, ideally making use of a saddle in the rim, and discharging into a tributary. Alternatively, a 'fuse-plug' emergency spillway might be introduced, in the form of an embankment across a saddle in the rim, which is designed to wash away if the water in the reservoir exceeds a certain level, during an extreme flood of low probability (see Fig. 7.1). This may be the level of the crest of the fuse-plug which is designed to retain this water level but, once it is overtopped, the downstream face and shoulder are designed to wash away. Discharge through the eroded gap in the rim then supplements discharge from the permanent spillway. To limit the extent of erosion and to determine the extent of the emergency spillway channel, the foundations and abutments of the fuse-plug may need to be reinforced. After the flood has subsided, it may be necessary for the fuse-plug to be re-built to restore storage for flood surcharge and freeboard (as discussed in Chapter 7).

(b) Strengthening of dams to allow raising or to counteract overstressing or cracking, may be achieved by buttressing the structure, or reinforcing it with post-tensioned cable anchorages. An informative case history in the case of a concrete gravity dam, has been described and summarized in Chapter 1 [12] and the general arrangement of the dam is illustrated in Fig. 17.1). Transverse joints in the dam had become solidified due to the deposition of calcium carbonates, and overstressing; consequent cracking and leakage occurred, due to restraint of longitudinal expansion. A

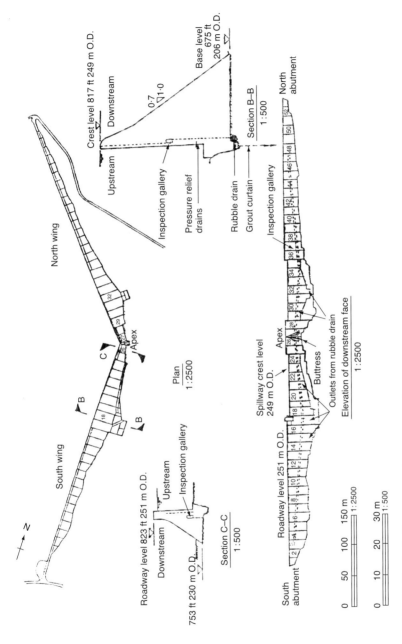

Fig. 17.1. Mullardoch Dam, general arrangement

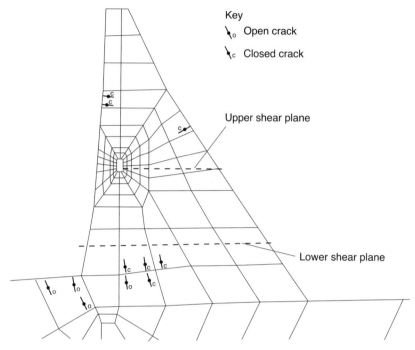

Fig. 17.2. Crack pattern in body of Mullardoch Dam after load step 5

solution was developed using a three-dimensional finite element model, which included representation of cracking emanating from a gallery through the length of the dam (Figs 17.2 and 17.3). The model compared the alternatives of cutting a transverse expansion slot or slots, or post-tensioning the central blocks of the dam to maintain the upstream face of the dam under compression. It was found that there would be unacceptable stress concentrations under transverse slots, and the solution chosen was to post-tension a central section of the dam, where the flanks intersected at an angle, with 26 semi-vertical tendons, each with an ultimate tensile strength of 11 100 kN. The cables were designed with the facility of monitoring and altering their loads and were provided with 'double corrosion protection' to allow extremely long life.

(c) Buttressing or raising of dams needs to be planned and designed with particular care because of the problems of obtaining compatibility and integrated action between the existing and the new sections. Considerations, in particular, are the thermal effects of the development of heat of hydration and subsequent cooling of new concrete placed against mature concrete, or the effects of

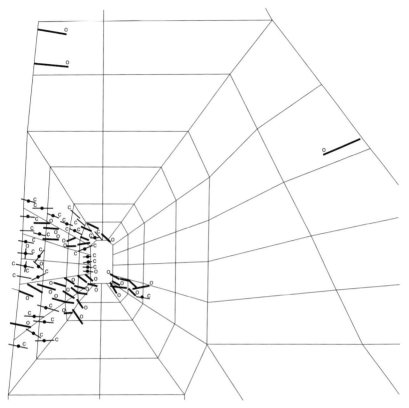

Fig. 17.3. Crack pattern around Mullardoch Dam inspection gallery after load step 5. Key as in Fig. 17.2

placing fill subject to settlement and consolidation against fill which is already well consolidated, and which restrains the consolidation of the new fill.

References
[1] *Interim Report of the Committee on Floods in Relation to Reservoir Practice,* (Reservoir Act 1930) Institution of Civil Engineers, ICE, London, 1933.
[2] *Interim Report of the Committee on Floods in Relation to Reservoir Practice: with additional data on floods recorded in the British Isles between 1932–1957* (Reservoirs Act 1930) Institution of Civil Engineers, ICE, London, 1960.
[3] Committee on Floods in the UK., Institute of Hydrology, UK, and National Environmental Research Council (NERC), *Flood Studies* — five volumes, 1975, reprinted 1978.
[4] *Floods and Engineering safety: An Engineering Guide,* Institution of Civil Engineers, 1st edition, 1978 and 3rd edition, Thomas Telford, London, 1996.
[5] The Reservoirs Act 1975, HMSO, London.

[6] Johnson, T. A., Charles, J. A. and Todd, P. (1990). *An Engineering Guide to the Safety of Embankment Dams in the United Kingdom,* the Building Research Establishment (BRE), 2nd edition, 1999.
[7] Charles, J. A., Abbiss, C. R., Gosschalk, E. M. and Hinks, J. L. (1991) *An Engineering Guide to Seismic Risk to Dams in the United Kingdom,* the Building Research Establishment (BRE), BRE, Garston (revised edition 1999).
[8] Construction Industry Research and Information Association (CIRIA). An Engineering Guide to the safety of concrete and masonry dam structures — Report No. 148, 1996, CIRIA, London.
[9] A Guide to the Reservoirs Act 1975. The Institution of Civil Engineers, for the DETR, i–viii and 1–209, Thomas Telford, London, 2000.
[10] A Guide to the Reservoirs Act *1975.* The Institution of Civil Engineers, for the DETR, part B 10(1) and 10(9), Thomas Telford, London, 2000.
[11] Technical guidelines for the appointment of panels of experts, and other provisions for the review of the design, construction, maintenance and safety of dams, (1983) Agricultural and Rural Development Department of the World.
[12] Gosschalk *et al.* Overcoming the build-up of stresses, cracking and leakage in the Mullardoch Dam, Scotland. 17th Congress of ICOLD, Vienna, 1991, Paper Q.65, R.26, pp. 475–498.

Further reading

Kennedy, N. F., Owens, C. L. and Reader, R. A. (1996). *Engineering Guide to the Safety of Concrete and Masonry Dam Structures in the UK.* CIRIA Report No. 148, 172.

18. Operation and maintenance, monitoring and inspection

General

It can be deduced from previous chapters that the safety and success of reservoirs depend on a series of outcomes which stem from sound planning and design. The final outcomes must be regular and comprehensive maintenance and correct and careful operation. The basis of these should be the final construction report (which the engineer responsible should prepare on completion of construction and commissioning), and the operating and maintenance (O & M) instructions, which should accompany the final report or follow it as soon as possible. With the concentration of attention on completion and commissioning and on the demobilization of both the engineer's and the contractor's staff, there is a risk that these reports may not be prepared with the thoroughness which is necessary. Any such neglect must be conscientiously avoided, but the preparation of the reports is quite a major undertaking, requiring meticulous attention, and is best carried out by persons thoroughly familiar with the relevant aspects of design and construction, with editing and co-ordination by the chief engineer.

Final construction report

The final construction report on a reservoir project should be prepared with the objective of making it a complete and sufficient record for future reference. Such a record might be required for the purposes of maintenance, modifications (including, for example, raising of a dam or strengthening it if this is ever found wanted), remedial works (for example dealing with cracks or countering unexpected leakage, settlement or deflections), providing a new discharge facility or spillway, and so on. It should therefore include at least the following information but consistent with this, it should be made as concise and readable as possible:

(a) a project description
(b) a list of the design parameters
(c) a copy of the technical specification
(d) a record of the hydraulic data relevant to the design
(e) geological and geotechnical data, including drawings, showing the results of surveys and investigations (including boreholes, with their locations) and results of tests
(f) a description of the design methods used and the principal results
(g) a copy of the key final as-constructed drawings, i.e. a set similar in content to those necessary as a basis for construction, but not necessarily including the less significant details. The location, bar sizes and centres and type of all reinforcement should be shown but not necessarily reinforcement details. Details of instrumentation installed should be included. The drawings might be marked-up design drawings and not necessarily be newly prepared
(h) a summary of the results of tests on materials used for construction
(i) a summary of grouting carried out, with locations, inclinations and depths of holes, quantities of cement injected, the relevant pressures applied and the results of water pressure or other testing
(j) a summary of the principal quantities of work carried out, with unit rates and costs
(k) a progress report on construction, including a comparison with programme
(l) a summary of additional (unforeseen) costs incurred, with explanations and comments
(m) a list of the principal authorities, engineers, architects, contractors, manufacturers and suppliers involved in design and construction, with addresses and names of their principals.

Operating and maintenance instructions

The primary actions for the production of O & M instructions start at quite an early stage: the design of the reservoir project should be such as to facilitate operation and maintenance, so that these procedures can be carried out safely and easily. Access routes, underground or within structures, should be well ventilated, and reasonably easy to negotiate by the staff likely to have to use them. Suitable routes should be planned for access to instrumentation, and for control, operation and maintenance of all E & M equipment and its removal, if eventually necessary. The reservoir, and particularly electrical and mechanical equipment and their controls, should be made secure against malicious entry. Manholes, shafts, handrails, ladders and hoists should conform with safety regulations and good practice, supported by the necessary inspection reports and (in the case of hoists and bulkhead gates) inspection and test reports. Adequate passing places in galleries and rest platforms (e.g. on ladders) should be provided. Clearances in shafts and galleries should be sufficient for the passage of personnel and equipment, which might include in some cases, for example, drilling equipment for further grouting to improve watertightness. The tops of shafts that are not covered by manholes (surge shafts for example) should be covered by removable frameworks and grills or mesh, to prevent careless or malicious dropping in of objects, and to prevent the entry of rocks falling from above, which might be carried by water to pumps or turbines. Drainage manholes should be provided at the ends of lengths of drains, suitable for clearing out the pipes or conduits by rodding or reaming. The works should be designed to divert sediment from entering intakes or conduits where it could cause blockages or damage.

The second type of early action required is to oblige manufacturers and suppliers of E & M equipment to supply, say, six copies of instructions for operation and maintenance of the equipment under their supply, related to the use for which the equipment is required. These should include all relevant details, including lubricating points, lubricants and frequency of lubrication required, precautions of any kind necessary in the use and maintenance of the equipment, paint specifications, parts lists sufficient for ordering and so on. The original orders for the equipment should include supply of sufficient spare parts expected to be required in the first, say, one year of operation, as recommended by the supplier, added to or revised by the Engineer, and approved by the Employer. One copy each of the O & M instructions supplied by suppliers and manufacturers should be kept in loose leaf binders or box files, as one or more volumes in the final O & M instructions for the project.

The preparation of the O & M instructions should have in view the manner in which they will need to be used. There may be a general section helpful for general management, followed by separate sections for each of the main components of the work, so that extracts may be separated and issued which are complete and relevant for the particular key members of staff who will have need or use for them.

The instructions should incorporate a chart showing, for each category of components of the works or in some cases for each component, the following guidelines:

(a) types of inspection required and frequency of each type
(b) frequency of lubrication required and details of lubricants (the same lubricant used for several items can be given an identifying letter or code)

(c) types of maintenance and cleaning required and frequency of each type, including renewal of corrosion protection
(d) comments on each item, including points to watch out for.

Monitoring and inspection
With the appointment of a Supervising Engineer (SupE) to be responsible for the day-to-day oversight of matters affecting the safety of a reservoir, the SupE becomes responsible for continuing surveillance and for deciding when it is necessary for inspections to be carried out, and to recommend inspections by the Inspecting Engineer. Key times are normally before the onset of the season most prone to floods, to ensure the proper preparedness of the dam for floods, and afterwards, to determine that the dam has behaved in accordance with the design, and that no unfavourable developments have occurred. Inspections should be carried out if unusual conditions or incidents have been reported which gave cause for concern to watchmen, visitors or others. ICOLD Bulletin 62 [1] recommends that inspection of a dam should be carried out if an earthquake of Richter magnitude 4 or greater has occurred with its epicentre within a 25 km radius of the dam, magnitude 5 or greater within a 50 km radius, 6 or greater within an 80 km radius and 7 or greater within a 125 km radius. Recommendations are given for the inspection and for any follow-up action which may be needed. The main tools of use during an inspection include the following.

(a) Review of reports on previous inspections and their implications for aspects to be observed, including review of the action which has been taken on recommendations in previous reports.
(b) Review of the records of monitoring since the last inspection. If thought necessary, special monitoring for the purposes of the inspection should be undertaken. Monitoring should include recording reservoir levels and observations of leakage. It may include:
  (i) measurements of settlement along the crest and, for embankment dams, within the dam
  (ii) measurements of deflection of the crest, perhaps currently best carried out by laser
  (iii) measurements of displacement and deformation, by geodetic survey of permanent survey points
  (iv) measurements of displacement and rotation by pendula, possibly 'inverted pendula', anchored at depth in the foundation, to provide a datum not subject to displacement, with the free end of the wire attached to a float, in a container of oil at a convenient high level

(v) measurements of rotation (and displacement) by inclinometer, within or on the dam
(vi) measurements of strain and/or extension within the dam, by vibrating wire or acoustic strain gauges, or by electrical resistivity gauges. (The latter, however, are not so reliable, being affected by the resistance and hence by the temperature of the connecting cable)
(vii) direct measurements of stress in concrete dams, the best known means being the Carlson stress gauge
(viii) cell pressure measurements in embankment dams
(ix) temperature measurements of the atmosphere and water and within concrete dams
(x) uplift and pore pressure measurements. These may be simple measurements of water level in standpipes, or by remote reading hydraulic or acoustic piezometers, which are more sensitive and responsive
(xi) measurements of seismic motion, by strong motion accelerographs (recommended at foundation level and crest level for dams in seismic categories III and IV, if more than 45 m high).

Measurements that are unusually large, or have moved outside a range anticipated from the design, or which indicate an unfavourable continuing trend, should be given special attention, and satisfactory explanations for them must be sought and, where not found, studies and action must be put in hand with appropriate urgency. Computerized automatic recording and monitoring systems should no doubt flag or signal measurements that are outside satisfactory limits set from the design.

(c) Observations by the inspector who should be looking out for:
　(i) reed growth, indicating possible seepage
　(ii) signs of leakage, especially new leakage. If emerging beside a culvert and if discoloured by suspended matter, it could be a sign of internal erosion of fill or of foundations
　(iii) irregularity in the lines of crest wall or kerbs or facings, which may indicate settlement or deflection or displacement
　(iv) cracks in either concrete, masonry or embankment dams, which may indicate excessive settlement or deformations or overstressing; in concrete dams, they could be a sign of alkali silicate reaction (ASR) which causes the concrete to expand; in fill dams, they could indicate slip-failure or differential settlement

(v) erosion, especially of the crest, and at the contact of the downstream face with the abutments — a possible result of overtopping — and erosion of surfaces subject to high-velocity flow, noting possible signs of cavitation
(vi) damage to the upstream face by wave action.
(d) Test of operation of gates and valves, noting if operation and controls operate correctly — irregular operation might be a sign of displacement of the guides or controls.
(e) Discussion and questioning with resident staff responsible for operation and maintenance.

Training and staffing

There are difficulties in staffing O & M requirements at reservoir sites, which are frequently remote and therefore present constraints for families. Provision should therefore be made for comfortable living accommodation and amenities. Remuneration needs to be attractive, both to attract suitable staff and to retain them. Clearly staff numbers have to be minimized for cost reasons, and this points to the maximum use of automated and automatic facilities, in so far as these can be relied upon. Even on a large project, in the limit, operation and management can be centred in a control centre with a very small number of staff, using transmitted data to be processed by computer, and leading to output which provides information on operation and performance, issues warnings of action needed and transmits signals which initiate controls. Printouts can be made automatically of whatever records are required. However, lengthy transmission distances entail their own problems and risks, for example line problems (including lightning, surges, wind, or animal or vegetable interference or radio interference). For conveyance of sensitive signals for control of equipment (the operation of which is critical for safety or other reasons), it may be necessary for the control processors to be located close to the equipment. Thus it will be necessary to have a minimum number of staff at, or available to visit, all significant components of the reservoir project, to observe correct operation and, if necessary, to take rapid action in case of incidents or problems. It follows that provision for stand-by and relief staff needs to be carefully planned and made to cover possible expected or unexpected absences, perhaps due to sickness or holidays.

Training of O & M staff is best initiated at an early stage in project implementation, so that staff who will be willing to continue with O & M on a semi-permanent basis after completion can be selected, and key persons can be sent to the works of manufacturers of the important E & M equipment for a period or periods of training.

These staff should receive further training by the manufacturers' commissioning specialists during commissioning of the equipment on site and by the civil engineers in charge where relevant, during construction. Provision for this should be made if possible in the tender documents and in the contracts for supply of the equipment, so that the arrangements and need for them are not left in any doubt.

Similarly, it may be desirable for selected members of future O & M staff to receive training with the consulting engineers, both civil and E & M. It can be of value, for example, for them to work with the consultants both on the site, during supervision of construction and installation of equipment, and in their offices, on preparation of designs, specifications and tender documents. In conjunction with this, it might be arranged for them to make visits to accessible relevant working projects, and to attend relevant conferences, talks and short courses that might be available at technical colleges or elsewhere. In doing so they should be expected to make a constructive and useful contribution to the work the consultants have in hand. It must be realized, however, that the training costs money — because one or more of the consultants' staff have to be on hand for most of the time, to accompany, brief and instruct the trainees and there is the cost of their needs for accommodation, subsistence, transport, etc. Their work inputs for a short period during training are not likely to be very productive because of the time they must spend on absorbing requirements, being briefed and possibly handing over on completion. It is therefore advisable to make plans and provisions for these costs early on, so that they are not overlooked and do not come as an unwelcome surprise.

It has to be borne in mind and provided against where possible, that persons who receive training of the kind required (which may be expensive and some of which may indeed be carried out overseas), once trained, may resign or leave to take other jobs more to their liking and perhaps better paid. This may well pose difficult problems in their replacement. It is therefore advisable to make the training subject to a written commitment (perhaps a contract, if possible) with the beneficiaries to become O & M staff on the terms and conditions that can be offered for a minimum period of a number of years, perhaps five, but certainly not less than two.

Potential unforeseen human, mechanical and electronic problems

Of course, the objectives of most of the work, including planning, design, supervision of construction, inspections, operation and maintenance, include that of trying to avoid unforeseen problems.

This might seem to be a contradiction in terms but in reality is not, because many problems which would otherwise be 'unforeseen,' can be foreseen by thorough investigations, sound design and good quality assurance throughout. There are, however, human errors, computer errors, breakdowns of plant and other problems which occur despite what seem to be the utmost precautions. For example, there have been errors in both the manual and computer-controlled operation of spillway gates, which have led to floods more severe and dangerous than would have occurred in nature. In one case, designs which were meticulously prepared by a very eminent European firm, for carefully supervised installation of electronic control of large radial gates, led to repeated failure and unsatisfactory operation, aggravated by delays in supply and delivery of replacement parts, and delays due to unavailability of the one or two specialist staff of the suppliers who were trained and qualified to rectify the faults.

Difficulties were magnified because the site was overseas, with some resulting delays in communication and travel. The result was that the local staff responsible for operation and maintenance lost confidence in the automated operation of the gates and instead of continuous automatic operation, established 24-hour manual attendance for control of the gates. This is a case where gate operation was likely to be required only infrequently when floods occur (though, being a 'flashy' river, not easily predictably), so continual attendance was akin to guard duty and very tedious, unless other activities could usefully be fitted in.

Another case was the occurrence of a flood in the middle of the night, above a minor barrage on quite a large river. Indeed, floods have a nasty habit of occurring in the middle of the night. There was a storm and main power supplies and telecommunications broke down. An assistant engineer was on duty, who knew that the radial gates ought to be opened. This the assistant managed to do, partly by manual operation but also by managing to start a stand-by diesel generator. Manual operation of heavy hydraulic gates is extremely laborious and time consuming and hence slow. It usually needs more than one person to operate a hoist. However, by prodigious efforts, the assistant managed to open the gates fully. It might have been better had the assistant not been able to do so, because the gates released such a flood that a cofferdam in the river downstream, and plant belonging to a contractor working on a diversion tunnel project, were washed away. This dramatically illustrates the necessity to ensure that the sudden opening of flood gates, when water levels upstream are high, must not be allowed to release a flood greater than one that would have occurred naturally in their absence.

Investigation and solution of leakage problems observed during inspections

The subject of tunnelling was discussed at some length in Chapter 10 and a case of leakage into underground excavations was described. In Chapters 1 and 17, the subject of remedial works was raised and reference was made to the case of a concrete gravity dam (48 m high) which was strengthened by high-capacity cable anchorages, to counteract overstressing and cracking. Further details will be given about this case now and then about another case in the Author's experience. Time is being devoted to the issue of leakage, because it is a problem which causes great concern if and when it occurs.

The problem at the concrete dam was first observed by a water man when he was making routine observations of water levels, temperatures and 'uplift' levels in standpipes, in a longitudinal gallery inside the dam (Fig. 17.1). He found that leakage into the gallery had increased in a short time from about 560 l/hr to about 18 750 l/hr. Pre-existing, nearly horizontal, cracks in the gallery walls, were found to have extended and opened from a maximum of 1 to 1·5 mm by up to about 1 mm; this was mainly evident in the two blocks extending about 15 m on either side of the central block of the dam, where the two flanks meet at an angle of 140° (pointing downstream). The increase in leakage caused the management considerable concern, because it was not known whether or not it could be a sign of much worse developments to follow; immediate steps were taken to lower the reservoir level, to carry out investigations and to recommend action. The concern was underlined because some ten years previously, precise surveys using Electronic Distance Measurement (EDM) had detected a slight tendency for the central apex of the dam to move in a downstream direction.

One of the early conclusions was that there was a need to reduce uplift pressures within the dam and at the interface between the dam and the rock foundations, where pressures were high because internal drains were blocked, and effective low-level outlets from the drainage system were lacking in the design. Additional low-level relief holes were drilled from the downstream face, and blockages in the internal drainage system were cleared by high-pressure jetting and flushing. The high internal uplift pressures immediately dropped, thus improving stability of the dam. Stability analyses, including two-dimensional, and linear and non-linear three-dimensional finite element studies, led to the conclusion that the cracking was accounted for by the seizing up of transverse expansion joints in the dam, coupled with thermal expansion, and geophysical inward movement (at times) of the valley sides, aggravated by the undesirably high internal uplift pressures which had occurred. Together these were capable of generating longitudinal compressive thrusts within the dam, which could result in vertical tensions sufficient to cause the cracking observed in the dam at its apex. The solution applied and proving satisfactory, was to post-tension the central section of the dam at the apex using 26 cable anchorages, each with an ultimate tensile strength of 11 100 kN. In 1995, five years after installation, the loads in the cables were monitored and topped up. The cost and the time required to investigate the problem, design and carry out the remedial works, were substantial but the dam had been in service satisfactorily for some 35 years when the problem demanded remedial action, and the action taken is expected to lengthen the life of the dam to as long as can be foreseen.

The other case to be mentioned here is one of excessive leakage which occurred at a rockfill dam overseas shortly after its completion. The dam is some 70 m high and has an upstream asphalt membrane. The leakage was observed from toe drains on the downstream side of the dam, mainly from the left side of central discharge culverts. In this case too, of course, investigations were immediately put in hand. An examination of the membrane with a sound transmitter, in an air bubble, traversing the membrane on a carriage suspended from the crest while listening for leakage noises, produced rather negative results. An inspection by a diver was undertaken to examine the possibility of cracks due to differential settlement in the asphalt membrane beside the intake to the central discharge culverts, where leakage noises had been detected. Some cracks were indeed found, but sealing of these cracks (underwater) produced negligible reduction in leakage measurements.

Depositing common salt on the bed of the reservoir towards the left abutment did, however, produce a marked response in increased conductivity of the leakage water. It was concluded that water paths must exist through the grout curtain in that area, which grouting had failed to seal. The best known specialists were consulted, and detailed and refined investigations were proposed to locate the leakage paths and to devise an economic solution to the problem. However, at this stage local concerns and political pressures had mounted to keep the reservoir in operation and to ensure safety without emptying the reservoir. There was also concern about the cost and the length of time that was being taken for investigations. A foreign consultant claimed that the problem could be solved by a programme of grouting, from above water level, without further investigations. The offer was accepted, but possibly its only merit was that everyone could see that something was being done to solve the problem. After some two years of work, this grouting had not had the desired effect.

The next step was to unload large quantities of impermeable soil over the crest of the dam into the reservoir, hoping that the soil would be drawn into the leakage paths and would seal the leak(s) — a method which had been considered previously but dismissed on the grounds that it might effect only a temporary remedy and meanwhile would obscure the leakage path(s) from detection. It appears, however, that this method achieved a satisfactory long-term reduction in leakage.

This case draws attention to the possible advantages of installing a grouting gallery at the top of the grout curtain or cut-off trench under a dam whilst it is being constructed. This has the merit of allowing testing and grouting to proceed while the dam is under construction on top, and to allow additional testing and grouting to be undertaken, if found necessary, at low level after the reservoir has filled. The disadvantage, of course, is that the gallery incurs additional initial cost. It might accelerate construction by isolating the grouting operations from other work, or it might delay construction by the introduction of the additional feature to be built (namely the gallery), so each case needs to be considered on its merits.

Other problems

In many cases, even in the UK, the data that would be required for new dams are simply not available for old dams, unless obtained at considerable cost. Dam owners, particularly, can easily take the view that, because a dam has served its purpose for many years without attention being drawn to any apparently serious signs of distress, it should continue to remain satisfactory without much expense. It has been demonstrated that this is not a safe assumption and inspecting engineers should approach it questioningly.

A case to illustrate this is a dam in China, which was built in 1959/ 60. It was constructed with a masonry spillway of unknown strength and soundness, with a mortar facing on the downstream face. Leakages were observed at more than ten places on the downstream face of the spillway. The long flanks of the dam were formed by earthfill embankment. The height of the masonry spillway was about 15 m. In view of a proposal to raise the dam, and due to a lack of data, site investigations were put in hand. These found that the dam was founded on a stratum of a kind of marl (soft, weak rock), which was supporting the existing dam with a factor of safety of less than one. It was not apparent why failure had not occurred in the 31 years since construction. Shear failure (sliding) in the foundations would have been most likely. As a result, it was decided that it was not feasible to raise the dam, but a new dam would have to be constructed to the height required, on a satisfactory foundation downstream.

To close on a more encouraging note, over a period of about 25 years the Author was responsible for inspecting a reinforced concrete buttress dam in Malaysia. This had been designed by British engineers, built by a British contractor and was completed in 1930. Detailed design drawings were available. The crest was raised by 1 m by the Japanese occupying the country during the war. Horizontal and vertical cracks were observed in the buttresses in the 1950s. They were monitored, and their number, lengths and widths progressively increased. The first inspection by the Author was in 1969, and studies (including finite element analyses), site investigations and the installation of instrumentation were progressively undertaken but without leading to a definite explanation for the cracking. At one time it appeared that progressive development of the cracking, if it continued, would lead to break-up of the dam. Proposals were prepared for strengthening the dam (by infilling the voids between alternate pairs of buttresses with concrete). Nevertheless, by 1984, increasingly refined instrumental monitoring demonstrated that the rates of change of both intensities of stress in the dam and of deformations were reducing and values were not likely to reach unacceptable limits. It was concluded that monitoring should continue but, in the absence of unexpected developments, the next inspection by a qualified engineer would not be necessary for another ten years.

The Author concluded that the cracking was most probably due to thermal effects, subject to restraints. He believes that, once cracks open, particles become separated and lodged in the cracks, preventing their complete closure. On the next thermal cycle, the cracks would be likely to open further and the process would continue, until perhaps at some stage, opening and closing of cracks would cease to occur progressively.

The Author believes that this demonstrates the need for an analytical but pragmatic approach: one must be prepared for the worst and safety must be maintained, but one should not jump unnecessarily to alarmist conclusions.

Reference
[1] International Commission on Large Dams (ICOLD), Bulletin 62. Inspection of dam following earthquake guidelines. Paris, 1988, 69 pages.

Index

Page numbers in *italics* refer to illustrations, tables and diagrams.

accerographs 27, *155*
access routes 306–307
adits
  hydrostatic testing 57
  trial 46, 53
aerial surveys
  geomorphological data 47
  topographical data 25
agriculture *see* irrigation projects
air entrainment 107, 215
Alimak raise climbers 196–197, *197*
alkeli silicate reaction [ASR] 309
alternative power supplies 18, 35, 224, 227
Ambursen dams 36
amenity uses, reservoirs 17, 19
Aswan High Dam, Egypt 222

Beard's Shortage Index 268
bellmouth ['morning glory'] spillways 107
Blackbrook Dam, England, earthquake damage 156–157, *158*
boreholes
  drilling 49, 52
  feasibility studies 45
  hydraulic fracturing tests 55–56, *56*, 57–58
  Lugeon values 55
  piezometers 46, 59–61, *60*
  rock stress measurement 57–58, *58*
borrow pits 49
bridges, spillway 106
Buckingham's method, mathematical modelling 73
bulb turbines *203*, 204–205
bulkhead gates, release by pressure equalisation 121
butterfly valves 215
buttress dams 36
  cracking 316
  diamond headed 37, *38–39*
    earthquake damage 157–158, *159*
  earthquake damage *145*, 151–153, *152*
  load spreading 37
  round headed 37
bypass valves, hydraulic gates 213

capacity factor, hydropower 251
catchment areas
  see also rainfall-runoff
  computer modelling 100

definitions 83
droughts, probability 97–99
elevation curves 29, *29*
floods
  data-based estimates 86–87, 89–91, 92–94
  probability-based estimates 94–97, *95, 96*
  geomorphological surveys 47, *48*
  hydrological surveys 83–84
  inflow, maximum 89–91
  maximising 23
  modifications, artificial 84
  subterranean losses 83–84
cavitation, slipways, energy dispersal 112, 115
channels, river diversion 119
Chile, earthquakes, dam damage 138–140, *138, 140*
civil engineers
  communication skills 9–10
  construction engineers 292–293
  inspecting engineers 293–294
  job opportunities 4–5
  overseas-based 5, 6
  qualified 292, 295–296
  supervising engineers 291, 294–295
  support specialists 8
  UK-based 5–6
Clywedog buttress dam, Wales 219, *221*
cofferdams
  incomplete dams as 120
  river diversion 118–120
  safety, floods 117
commissioning 20
communities, displacement 220, *232*, 234
compaction tests 49, 54–55
completion dates 18
computational fluid dynamics [CFD] 67–68
computers 8–9
  software
    catchment hydrology 100
    costing 241–242
    flood allowances 91
    hydraulic modelling 68–71
concrete
  allowable tensile stresses 154–156
  pozzolanic content 37
  rates of production 40, *40*
  roller compacted [RCC] 37, 40
  spillways 106–107

concrete arch dams 27, 36
  earthquake damage *145*, 153
  foundations 37
  multiple 36
concrete dams
  alkeli silicate reaction [ASR] 309
  cracking 309, 316
  earthquake damage *145*, 160
    peak tensile stresses 151, *151*
    resistance to 160-161, *161*
  hollow buttress 36
  mass 36
  rates of building 40, *41*
  spillways, floods 42
concrete gravity dams 36, *161*
  earthquake damage 153-157, *158*
  gallery leakage 313-315
conduits
  see also tunnels
  free-flowing enclosed 174
  open
    construction 167-168
    freeboard 168
    routing 174
  power losses in 250-251
  pressure 168
construction engineers 292-293
construction materials
  availability 37
  concrete, specifications 37, 40
construction stage 20
  certification 293
  final reports 305-306
  floods
    insurance 118
    probability 117-118
  infrastructure 243
  modifications, recording 305
  temporary works 118, 243
consultancies 4-5, 6
contingency planning 19-20
corrosion
  hydraulic gates/valves 217-218
  inspections for 309-310
  pipelines 173-174
cost-benefit analysis 253
costing
  accuracy 243
  adjustments 35, 244, 254
  allowances 43
  alternative power sources 35
  computer software 241-242
  construction infrastructure 243
  contingencies 246-247, 254
  dam heights 43
  feasibility stage 61, 242-243
  financing agencies 243
  fossil fuels 244, 254
  indices *246*
  insurances 43
  interest allowances 244
  internal rate of return [IRR] 248, 253
  margins of capacity 21, 35
  mark-ups 244, *245*, 246
  materials *245*, 246-247

  multiple contractors 43
  net present values [NPV] 253
  operation and maintenance [O and M] 249
  opportunity 247-248
  pay back period 248
  prefeasibility stage 61, 242-243
  present value 247
  sensitivity analysis 254
  shadow prices 248
  sinking fund 248-249
  taxation 243-244
  temporary constructions 43, 243
  test discount rate 252
cranes, travelling, loading bays 201, 208-210, *208*, *209*
cross-flow turbines *204*, 206

dams
  see also World Commission on Dams
  choice of 42
  concrete
    arch *27*, 36, 37
    hollow buttress 36
    mass 36
    multiple arch 36
  embankment
    earthfill 35, *36*
    rockfill 35, 40, *41*, 42
  environmental impact, assessments 230-231, *231-232*, 233-237
  failures 289, 290-291
    breaches 101, *102*, 103
    shear 26, 45-46, 315
  foundations 36-37
  height 34
    optimum 43
  large, need assessments 228-229
  natural loading allowances 101
  old, data lacking 315-316
  raising 302-303, 316
  sites, ideals 24
  strengthening 300, *301*, 302, *302*, *303*
dead storage 15, *16*
deformations, dynamic 58-59
Deriaz turbines 203
diamond headed buttress dams 37, *38-39*
dimensionless similarity parameters 73, 76
discharge outlets
  control gates 110
  energy dispersal 110
  low-level
    control 109-110
    diversion tunnels 109
    embankment dams 110
    screening 110
downstream displacement, by earthquakes 153
droughts
  probability 98
  probable worst [PWD] 97-98

earthfill dams 35, *36*
earthquakes
  core liquifaction 139, 143, *146*, 159-160
  damage 141-148, *142*, *143*, *146*, *147*
  foundations 36

earthquakes
  see also peak ground acceleration
  ancillary structures, damage 158–159
  attenuation 140–141, *141*
  Chile, dam damage 138–140, *138*, *140*
  concrete dams, damage *145*, 149, 151, *151*
  downstream displacement by 153
  earthfill dams
    core liquifaction 138, *146*, 159–160
    damage 141–148, *142*, *143*, *147*
  focal depths 133
    NW Europe *125*
    UK *125*
  ground displacements 134, *135*
  incidence
    UK 123–125, *123*, *124*, *125*
    world *137*, 138
  intensity, distance form epicentre 140–141, *141*
  maximum credible [MCE] 132
  Modified Mercalli scale 133
  prediction 161
  probability, UK 125–126
  reservoir induced [REI] 149, 157, 158, 222–223, 236
  Richter scale 132–133
  rockfill dams
    damage *144*, 148–149, 160
    resistance to 160, *160*
  seiches 143
ecosystems, destruction 221–222
electrical and mechanical [E and M] engineers, consultancies 199–200
electro-mechanical equipment
  see also hydraulic gates; turbines
  down time costs 249
  generators 207
  loading bays 201
    cranes 201, 208–210, *208*, *209*
  operation and maintenance instructions 307–308
  site capacity 200–201
  small-scale 273
  supply 200
  transformers 207–208
  transmission systems 210–211, *210*, *211*
embankment dams
  earthfill 35, *36*
  low-level discharge outlets 110
  rockfill 35
    membranes 40, *41*, 42
  spillways 42, 111
emergency preparedness plans [EPP] 136
emergency situations
  mechanical failures 311–312
  storms 312
empirical approaches, constraints 10–12
encrustation, hydraulic gates/valves 217
energy dispersal
  cavitation 112, 115
  discharge outlets 110
  spillways 111–112
    baffled 113–114, *114*
    ski [flip] bucket 112–113
    stepped 112, *112*
    stilling pools/basins 113–114, *113*, *114*
  valves 115

enforcement authorities 292, 297–299
environmental impact
  assessments 230–231, *231–232*, 233–237
  beneficial 231, *231–232*, 233, 237–238
  detrimental 231, *231–232*, 233
  displacement of persons 220, *232*, 234
  downstream flow *232*, 235
  evaluation 222
  international standards 227
  legislation 219
  migratory tracks *232*, 234–235
  natural resources *232*, 233–234
  new projects 19
  sedimentation rates 219–220, *221*
  soil salination 220
  weed growth 235
evotranspiration, potential 262–263

faults
  see also earthquakes; seismic risks
  leakages 24
  tunnels
    bridging 194, *195*
    realignment 192–194, *193*
    remediation options 188–189, *188*
    routing 174–175
    sealing strips 186, *187*
fauna, environmental impacts on *232*, 234–235
feasibility studies 20
  costings 61, 242–243, 247–248
  on-site investigations
    geological data 45–46
    geomorphological data 47–48
    rocks 54–57
    trial pits/trenches 48–49
financing agencies, costing 243
firm energy output, hydropower 252
firm power output, hydropower 251–252
fixed-bed modelling 79–80
Flood Estimation Handbook [FEH] 87–89
floods
  cofferdams 117
  construction stage, probability 117–118
  control 17
  design allowances, maximum 87
  discharge outlets, low-level 109–110
  frequency
    data-based estimates 86–87, 88, 92–94, *92*, *93*
    probability-based estimates 94–97, *95*, *96*
    UK 87–89
  probable maximum [PMF] 87, 90, *102*, *103–104*
  protection, Nile 222
  rainfall, maximum annual 96–97
  reservoir storage 15, *16*
    safety factors 88–89
  reservoirs
    computer modelling 91
    health and safety 88–89, 91
  spillway discharge 106–109
    potential 32–33, 86
  surge waves 86
flow simulation software 70–71
FLOWMASTER software 70

Forres project, Scotland, shaft excavation 196–197, *197*
fossil fuels, premiums 244
foundations
 earthfill dams 36–37
 unstable 315–316
Francis turbines 201–202, *202*
freeboard 16, *16*
Froude law of similarity 76, 79
full supply levels 15
fuse-plug spillways 110–111, *111*

gated crest spillways
gates
 bear trap [flap] 108–109
 dimensions 107
 drum 108
 Hydroplus 109
 radial [Tainter] 108
 vertical lift 108
 operation 108
 vibration 108
gates *see* hydraulic gates/valves
generators
 see also electro-mechanical equipment
 asynchronous/induction 207
 governors 207
 location *169*
 synchronous 207
geographical data, computerized 18
geological data
 appraisal 26
 field mapping 46
 landslides 26
 local knowledge 30
 new sites 18
 prefeasibility studies 30
 recording 306
 rocks
  quality indices 52–53
  void indices 53
 sub-surface investigations
  adits 46, 53
  boreholes 45, 46, 49, 52–53
  sampling 46
  trial pits 46
geomorphological data 46
 aerial surveys 47
 catchment areas 47, *48*
 maps *52*
geophysical surveys 46
 seismic methods 53–54
ground deformation tests 54–55, 57
ground displacements, earthquake-induced 134, *135*
grouting
 curtains 37
 galleries 315
 test 56–57
 tunnels 182, 186

Hegben dam, USA 143, *146*
hollow jet valves 216
Howell Bunger valves 215, *216*
Hsinfenkiang Dam, China, earthquake damage 157–158, *159*

hydraulic fracturing tests 55–56, *56*, 57–58
hydraulic gates/valves
 air entrainment 107, 215
 butterfly 215
 bypass valves 213, *214*
 control 213, *214*
 corrosion 217–218
 emergency 213, *214*
 encrustation 217
 energy dispersing 213, *214*
 flood discharge 213, *214*
 guard gates/stop logs 212–213
 hollow jet 216
 Howell Bunger 215, *216*
 specifications 212, 215
 spherical 216
 vertical lift 216
 vibration 108, 213, 215, 217–218
hydraulics
 computational modelling 67–68, 69
  software 68–71
 empirical basis 66–67
 mathematical analysis 67
 numerical modelling 65
 prediction-based 65
 theoretical analysis 65–66
hydrological data
 cyclical influences 28
 incorrect, consequences 63
 investigations
  analytical methods 64, 66–67
  computational models 64, 67–71
  past experience 63, 64–65
  physical models 64, 71–81
 new sites 18, 27–28
 outflow estimates 33–34, *33*
 rivers
  flow records 30–34, *31*, *33*
  mass curves 32–33, *33*
  profiles 28–29
hydropower 16
 alternatives 224, 272
 capacity factor 251
 demand profile 19
 efficiency estimates 34–35
 firm energy output 252
 firm power output 251–252
 generation potential 34–35, *34*
 generators, location *169*
 load factors 251, 275–276
 losses
  conduit intake/outlet 250–251
  transmission 249
 projected demands 225
 pumped storage 16–17
 reservoirs, design criteria 24
 small-scale
  advantages 272, 276–277
  classifications 271–272
  demand response 278–279, *278*
  economics 275, 277, 288
  evaluation studies 283, *284*–*286*, 287
  feasibility studies 281
  global spread 271
  grid connection 280, 282, *282*

infrastructure 274–275
location 279–280, 282–283
multiple 274–275
planning 272–274, 277–279
private 280, 281–282
Scotland 280
test discount rate 252
thermal power comparison 252
tunnels 168, *169*, 175
linings 185–186
water hammer, safety factor 177

impounding 15
certification for 120
earthquakes induced by 149, 157, 158
Indonesia, flood data 92–94, *92*, *93*, *95*
inspecting engineers 293–294
inspections
certificates 294
leakages
diagnoses 313–314
remedial actions 314–315
remedial work following 300, *301*, 302–303, *302*, *303*
supervising engineers [SupE] 308–310
insurance
construction stage, floods 118
costing for 43
internal rate of return [IRR] costing 248, 253
internet, uses of 13
Irrawaddy River Delta, Myanmar 10–12, *11*
irrigation projects
see also conduits
agricultural demands 225, 255–256
and aquifers 262
control, gates 266–267
crop yields 262, 266
distribution systems 263, 264
closed conduits 264
sprinklers 263–264
ecosystems destruction 221–222
effectiveness, measuring 269
efficiency, measuring 256, 269
forecasting requirements 267–268
inefficiencies 256
losses 256–257
minimising 258, 263, 264
planning
for efficiency 258, *259*, *260*, 261, 262
water requirements 261–262
potential evotranspiration 262–263
reservoirs 16, 23
soil data 18
soil salination by 220
storage, balancing 264–265
water shortage periods 268
water sources 265
ISIS FLOW software 69–70
Itaipu Dam, Brazil *38*

Kaplan turbines 202–203, *203*
Kotamale valley
Sri Lanka *48*, *52*
tunnelling 185–191, *186*, *187*, *188*, *190*

Koyna dam, India, earthquake damage 149, *150*, 151, *151*

lakes, Reservoirs Act [1975] 298–299
landslides
evidence of 26, *52*
reservoir induced 236
leakages
erosion by 37
faults 24
grout curtains 37, 315
increasing
diagnoses 313–314
remedial actions 314–315
minimising 226
sealing strips 186, *187*
tunnels 176, 185–187
least-cost analysis 253
legislation
see also Reservoirs Act [1975]
environmental impact 219
limestone, karstic characteristics 24
live storage 15, *16*
load factor, hydropower 251
loading bays
cranes 201, 208–210, *208*, *209*
generating equipment 201
losses
hydropower, conduit head 250–251
irrigation projects 256–257, 258
rainfall-runoff 83–84, *85*, 250
transmission systems 249
Lower Crystal Springs Dam, USA, earthquake damage 154–156, *156*, *157*

Manning's equation, slope scale distortion 79
margins of capacity 21, 35
mass concrete dams *see* concrete gravity dams
mathematical modelling 12
Buckingham's method 73
method of similitude 74
Rayleigh's method 73, 79–80
Mauvoisin Dam, France 26, *27*
maximum credible earthquakes [MCE] 132
mechanical and electrical [E and M] engineers, consultancies 199–200
Mersey Tunnel, England, faults 194, *195*
meteorological data
Beard's Shortage Index 268
irrigation projects 267–268
new sites 18
MICROFSR2 software 91
MIKE 11 software 70–71
minimum operating levels 15, *16*
modelling
computer software 68–71
dimensional analysis 72–73
dimensionless similarity parameters 73, 76
fixed-bed 79–80
free-surface flow 78–79
mathematical 12
Buckingham's method 73
method of similitude 74
Rayleigh's method 73, 79–80
movable-bed models 80

modelling (continued)
  prototype materials 77
  scale
    effects 81
    ratios 77–78
  sediments 80
  slope scale distortion, Manning's equation 79–80
  vertical scale distortion 79
  viscosity 78
Modified Mercalli scale 133
Monar Dam, Scotland 219, 220
Mullardoch Dam, Scotland, remedial work 300, 301, 302–303, 302, 303

natural resources, environmental impact 232, 233–234
net present values [NPV] costing 253
New Austrian Tunnelling Method [NATM] 180

open spillways
  advantages 106
  bellmouth 107
  bridges over 106
  crest erosion 106
  sealing layer 107
  surfacing 106–107
operation and maintenance [O & M]
  costs 249
  electro-mechanical equipment 307–308
  instructions, access routes 306–307
  staff 310–311
opportunity costing 247–248
overtopping
  see also spillways
  rockfill dams, erosion 120
  surge waves 102, 105

Pacoima Dam, USA, earthquake damage 153, 154, 155
pay back period costing 248
peak ground acceleration
  dam categories 131
  distance from epicentre 140–141, 141
  Koyna concrete dam 149
  UK 125–126, 127, 128, 131
peak tensile stresses, concrete dams, earthquakes 151, 151
Pelton turbines 204, 206
permeability tests 55
Philippines, flood data 92–94, 92, 93, 96, 96
piezometers, in boreholes 46, 59–61, 60
pipelines
  corrosion 173–174
  expansion 173
piping, fluid flow, software 70
planning
  contingencies 19–20
  effective 269
  irrigation projects 258, 259, 260, 261, 262
  margins of capacity 21
  principles 19–21
  sites, electro-mechanical equipment 200–201
  small-scale hydropower 272–274
  time schedules 20–21

precipitation
  flooding as function of 96–97
  irrigation projects, supply 265
  maximum, UK 91
  probable maximum [PMP] 87, 91, 104
prefeasibility studies
  construction materials, availability 37
  costings 35, 43, 61, 242–243
  environmental impact, assessments 230–231, 231–232, 233–237
  hydrology 30–35
  parameters 20, 29
  reconnaissance studies, team members 30
present value costing 247
probability
  Gumbel's extreme value distribution 94–95, 95
  Pearson distributions 94
  time series distributions 94
probable maximum floods [PMF] 87, 90, 102, 103–104
probable maximum precipitation [PMP] 87, 91, 104
probable worst droughts [PWD] 97–98
projects
  constraints 17–18
  continuity 7
  extant information 18–19
  monitoring panels 7
  programming work 19–21

rainfall see precipitation
rainfall-runoff
  assessment, time factor 85–86
  factor 85
  losses
    absorption 85
    evaporation 84
    subterranean 83–84
  maximum 89–91
raise-boring shaft excavation 195–196, 196
Rayleigh's method, mathematical modelling 73, 79–80
reservoirs
  aesthetic qualities 222, 235
  alterations to 295–296
  benefits, quantification 224
  functions 16–17
  health and safety, flooding assessments 88–89, 91
  impounding 15, 120
  induced earthquakes [RIE] 149, 157, 158, 222–223, 236
  landslips 236
  large, defined 290
  life estimates 221
  new, expert review 299–300
  pollution 235
  safety legislation see Reservoirs Act [1975]
  sediment deposition 84, 219–220, 221, 273–274
  service/balancing 15
  size as function of utility 224–225
  storage potential, estimates 34
  volcano proximity 236–237
  water chemistry 235–236
  weed growth 235

Reservoirs Act [1975]
    breaches of 294, 297
    cofferdams 117
    construction certification 293, 296
    construction engineers 292–293
    emergencies 298
    enforcement authorities 292, 297–299
    impounding commencement 120
    inspecting engineers 293–294
    inspection certificates 294
    lakes 298–299
    provisions 290
    records 296–298
    revision 299
    supervising engineers 291, 294–295
    undertakers 296–297
Reservoirs [Safety Provisions] Act [1930] 289, 290
Reynolds law of similarity 76–77
rivers
    diversion
        cofferdams 118–119
        completion 120–121
        incomplete dams 120
        tunnel/channel excavation 119
    downstream
        dam breaches 101, *102*, 103
        flow 86, *232*, 235
    fauna, migrant species *232*, 234–235
    flow duration curves 31–32, *31*
    flow forecasting, irrigation projects 268
    mass curves 33–34, *34*
    profiles 28–29
    reservoir outflow estimates 33–34, *33*
rockfill dams 35
    asphalt membranes 40, *41*, 42
    earthquake damage *144*, 148, 149, 160
        crest settlement 148–149
    earthquake resistant *160*
    overtopping, erosion by 120
rocks
    hydraulic fracturing tests 55–56, *56*
    Lugeon values 55
    quality designations [RQD] 52–53
    stress measurement 57–58, *58*
    void indices 53
rockslides 26
roller compacted concrete [RCC] 37, 40
    rates of production 40, *40*
    spillways 106–107
round headed buttress dams 37
runoff *see* rainfall-runoff

safety
    see also Reservoirs Act [1970]
    adits, ventilation 53
    evaluations 132
    reservoirs, flooding assessments 88–89, 91
    seismic activity
        hazard plans 136
        potential 131–132
samples 54
San Fernando Dams, USA, earthquake damage 143–144, *143*, *147*
San Francisco earthquake [1906] 141–142, *142*

sediments
    reservoir deposition
        assessments 84
        rates 219–220, *221*
    transport, modelling 80
Sefid Rud Buttress Dam, Iran, earthquake damage 151–153, *152*
seiches 143
    see also surge waves
seismic activity *see* earthquakes
seismic data
    accerographs 27
    new sites 18, 26–27
    UK 123–125, *123*, *124*
seismic risks
    see also earthquakes; peak ground acceleration
    concrete, allowable tensile stresses 154–156
    dams
        classifications *129*, *131*
        UK 128–129, *129*, *130*
    effects 131–132
    emergency preparedness plans [EPP] 136
    maximum credible earthquakes [MCE] 132
    shear-friction factors, interfaces 133
seismic surveys 53–54
    cross-hole 54
sensitivity analysis 254
shadow prices costing 248
shaft excavation
    Alimak raise climbers 196–197, *197*
    raise-boring 195–196, *196*
shear failures
    dams 26, 315
    detecting potential 45–46
shear strength, tests 55
shear-friction factors, interfaces 133
siltation indices *50*, 51
similarities
    dimensionless parameters 73, 76
    Froude law of 76, 79
    laws of 75, 76
    Reynolds law of 76–77
similitude, method of 74
sinking fund costing 248–249
site data, appraisal 25
site planning, electro-mechanical equipment 200–201
slope instability, investigations 26, 47–48
soils
    borrow pits 49
    erosion 219–220, *221*
    laboratory tests 49
    on-site testing 48–49
    salination 220
    sampling 48
    triaxial testing 49
spherical valves 216
spillways
    see also conduits; discharge outlets
    air entrainment 107
    baffled 114, *114*
    bellmouth 107
    concrete dams, floods 42
    cracking 107, 315
    embankment dams, floods 42

spillways (*continued*)
 emergency fuse-plugs 110–111, *111*, 300
 energy dispersal 111–115, *112*, *113*, *114*
 floods, discharge capacity 32–33, 86, 300
 gated crest 107, 108–109
 gates, operational planning 86
 interface erosion 42
 open 106
 overtopping
  side walls 105
  surge waves *102*, 105
 sealing layer 107
 smooth 113–114, *113*
 stepped 112–113, *112*, *114*
 surfaces
  concrete finish 106
  paving 106–107
stop logs 121–122, 212–213
straight-flow/Straflo turbines *204*, 205–206, *205*
supervising engineers [SupE]
 inspections 308–309
 observations 309–310
 statutory requirements 291, 294–295
 tests 310
surge waves
 see also seiches
 floods 86
 rockslides 26
 wind, allowances *102*, 105

taxation, local 243–244
temporary constructions, costing 43, 243
tenders 20
 documents, river diversion 118
test discount rate 252
theoretical approaches, constraints 10
topographical data
 appraisal 25
 mapping scales 25
 new sites 18
training, operation and maintenance staff 310–311
transformers 207–208
transmission systems
 cables 211
 losses 249
 voltages 210–211, *210*, *211*
transport
 access routes 306–307
 reservoir-based 237
 transformers 208
trial pits/trenches 46
 soil tests 48–49
tubular turbines *203*, 205
tunnels
 boring
  drill and blast method 173, 175
  machines [TBM] 173, 176, 197–198
 cracks
  grouting 182, 186, 192–193, *192*
  remediation options 188–189
  sealing strips 186, *187*
  steel penstock bypasses 188, *188*
 cross-sections *171*, *172*, 173, 182–183
 diameter, optimising 178

faults, unforeseen 192–193
flow velocities 178–179
hydropower 168, *169*, 185–191, *186*
linings
 cast in situ concrete 170, *171*, 179–180, 185–186
 precast concrete 170, 180
 reinforced concrete 170–171
 rock bolts 180
 shotcrete 170
 welded steel 171–172, *172*, 181–182, 185, 189–190
overburden depth 176
realignment 192–194, *193*
river diversion 119
rock 168–170
routing 174–176
 faults 174–175
shaft excavation
 Alimak raise climbers 196–197, *197*
 raise-boring 195–196, *196*
spoil disposal 182
stresses 176–177, *177*, 179, 186–188
supports 197
surges
 chambers 168, *169*
 safety factors 177
water drainage, gravity 192–193, *193*
turbines
 see also electro-mechanical equipment
 bulb *203*, 204–205
 cross-flow *204*, 206
 Deriaz 203
 Francis 201–202, *202*
 Kaplan 202–203, *203*
 Pelton *204*, 206
 pumps in reverse 206
 straight-flow/Straflo *204*, 205–206, *205*
 tubular *203*, 205
 turgo 206
turgo turbines 206

Vaiont Dam, Italy 26, *27*
valves *see* hydraulic gates/valves
vertical lift valves 216
vibration
 gated crest spillways 108
 hydraulic gates 213, 215, 217–218
Victoria Hydropower project, Sri Lanka
 faults 191–194, *191*, *192*, *193*, *194*
 shaft excavation 195–196, *196*
void indices, rocks 53
volcanoes, reservoirs close to 236–237

WASP [Wien Automatic System Planning] software 241–242
water supply, increasing demands 223–224, 229
weed growth, reservoirs 235
WIGPLAN software 242
Wimbleball Dam, England *39*
World Commission on Dams
 environmental considerations 227
 good practice guidelines 228–229
 mandate 226
 policy framework 228